Synthesis Lectures on Computer Science

The series publishes short books on general computer science topics that will appeal to advanced students, researchers, and practitioners in a variety of areas within computer science.

Andrew Sabot

Minimizing Data Movement and Parameter Count Across the Machine Learning Stack

Everything is a Matrix

Springer

Andrew Sabot
New York City, NY, USA

ISSN 1932-1228 ISSN 1932-1686 (electronic)
Synthesis Lectures on Computer Science
ISBN 978-3-032-23099-7 ISBN 978-3-032-23100-0 (eBook)
https://doi.org/10.1007/978-3-032-23100-0

This Springer imprint is published by the registered company Springer Nature Switzerland AG
The registered company address is: Gewerbestrasse 11, 6330 Cham, Switzerland

Dedicated to my grandparents, who were always supportive and proud of me and whatever I wanted to do.

Preface

The rapid ascent of machine learning has redefined the boundaries of what is computationally possible, from real-time natural language understanding to sophisticated computer vision. However, as the field of artificial intelligence continues to progress, the industry has reached a critical juncture: the sheer scale of modern models has outpaced the growth of the hardware required to run them. The "computational tax" of artificial intelligence is no longer just a technical hurdle; it is a sustainability and accessibility crisis.

This book is born out of the necessity to bridge the widening gap between algorithmic ambition and hardware reality. At its core, it argues that the next leap in AI will not come solely from larger datasets or deeper layers but from fundamental innovations in how we manage the matrix operations that underpin a substantial amount of modern workloads.

Scope and Origins

The work presented here originated from my doctoral research, where the goal was to rethink the fundamental units of optimization in scientific computing and neural networks. While many texts focus on the high-level architecture of AI, this book dives "under the hood" to explore the intersection of linear algebra, hardware utilization, and model compression.

Overview of the Book

The chapters are organized to guide the reader through a hierarchy of optimization strategies:

- **Optimal Data Movement:** We begin with the CAKE and mCAKE frameworks, which address the most common bottleneck in modern systems: memory bandwidth. By partitioning computations into optimally shaped blocks, we demonstrate how to unlock latent performance in existing hardware.

- **Exploiting Sparsity:** With the Rosko framework, we move into the realm of structured pruning and outer-product exploitation, showing that we can achieve significant speedups by simply not computing what does not need to be computed.
- **Model Compression and Distillation:** We explore the power of Singular Value Decomposition (SVD) and low-rank representations. These chapters provide a roadmap for creating "student" models that maintain the intelligence of their "teachers" while occupying a fraction of the memory footprint.
- **Privacy and Distributed Inference:** Finally, the book addresses the ethical and logistical challenges of cloud-based AI, offering a framework for splitting inference between local devices and the cloud to preserve user privacy.

Target Audience

This book is intended for researchers, high-performance computing (HPC) engineers, and graduate students who are looking for more than just a theoretical understanding of AI. It provides actionable techniques for those tasked with deploying large-scale models in resource-constrained environments. This book includes optimization strategies ranging from on-device edge computing to massive, energy-conscious data centers.

Previously Published Material

Portions of this monograph are adapted from prior publications and ongoing collaborations. Chapter 2 is based on work originally published as *CAKE: Matrix Multiplication Optimization* [1] and incorporates material from *MEMA: Memory-Efficient Matrix Algebra* [2]. Chapter 4 draws on work appearing in *Rosko: Row-skipping Outer-product Execution* [3]. Chapters 5 and 6 expand upon research first presented in *GASA: Generative Adaptive Sparse Approximation* [4].

Finally, Chaps. 3, 7, and 8 include material from internal working papers developed in collaboration with Vikas Natesh, Mark Ting, He-Yen Hsieh, and H.T. Kung.

References

1. H. T. Kung, Vikas Natesh, and Andrew Sabot. CAKE: Matrix multiplication using constant-bandwidth blocks. In *Proceedings of the International Conference for High Performance Computing, Networking, Storage and Analysis*, SC '21. Association for Computing Machinery, 2021.
2. Andrew Sabot, Vikas Natesh, H. T. Kung, and Wei-Te Ting. MEMA Runtime Framework: Minimizing External Memory Accesses for TinyML on Microcontrollers. In *Proceedings of the tinyML Research Symposium*. tinyML Foundation, San Francisco, CA, 2023. Extended abstract; also available as arXiv:2304.05544 [cs.LG]. https://arxiv.org/abs/2304.05544

3. Vikas Natesh, Andrew Sabot, H. T. Kung, and Mark Ting. Rosko: Row Skipping Outer Products for Sparse Matrix Multiplication Kernels, 2023. https://arxiv.org/abs/2307.03930.
4. H. T. Kung, and Andrew Sabot. GASA: Rank-Sliced Gather-Scatter Activations and Its Application to Sparsity-Preserving Parameter-Efficient Fine-Tuning. In *Proceedings of the IEEE International Symposium on Circuits and Systems (ISCAS)*. IEEE, London, United Kingdom, 2025.

New York City, NY, USA
January 2026

Andrew Sabot

Acknowledgments

This work is the culmination of a journey supported by many extraordinary people. I am deeply grateful to my PhD advisor, H.T. Kung, for the extraordinary time and wisdom he invested in my growth as a researcher. Our countless hours discussing challenging problems and their potential solutions have profoundly shaped the scientist I am today.

I would like to specifically thank my colleagues in the Kung Lab, particularly Mark Ting and Vikas Natesh, whose collaboration and support were instrumental to the development of the frameworks presented in this work. I also thank my committee members, Harry Lewis, James Mickens, and Gage Hills, for their invaluable guidance. Furthermore, I am grateful to Harvard University for providing the intellectual environment and resources that made this research possible.

I thank my friends and family for their unwavering support. Most importantly, I thank my parents for their daily encouragement, and my brother David Sabot for his ability to lift my spirits throughout this journey. I also owe a special thanks to my dog, Rory Sabot, who always knew exactly when it was time for a walk break. To everyone who touched this project—thank you for your belief in me.

Competing Interests The author declares the following as potential competing interests: The research presented in this monograph was supported in part by the Air Force Research Laboratory (AFRL) under award numbers FA8750-18-1-0112 and FA8750-22-1-0500, and Meta Platforms Technologies under award number A51540, and through various grants awarded to the Kung Lab at Harvard University.

Contents

1 Introduction

1.1 The Widening Gap

Machine learning and artificial intelligence (AI) have transformed not only our daily lives—from predictive text on smartphones to real-time language translation—but also entire industries, including healthcare diagnostics and large-scale climate modeling. At the core of these advances are deep neural networks, whose scale and complexity have surged rapidly in recent years. This relentless growth increasingly stresses computational hardware, creating bottlenecks not just in processing speed but also in memory bandwidth, energy efficiency, and data movement.

Deep learning models have delivered breakthroughs in areas such as computer vision and natural language processing. However, large models such as language and vision transformers [1,4], while increasingly powerful, require ever more compute resources. As Moore's Law slows and sustainability becomes paramount, a central challenge emerges: How can we continue to enable innovation and broader deployment of AI while working within stringent computational and energy constraints?

Even as models and their hardware evolve and diversify, one common thread remains: the matrix. Matrix operations, especially multiplication, decomposition, and reduction, form the computational backbone of nearly all machine learning tasks, regardless of whether the architecture is convolutional, recurrent, or transformer-based. In fact, matrix calculations can constitute over 99% of the workload in state-of-the-art models [2]. As a result, efficient matrix computation is not only foundational but also a model-agnostic problem; innovations in this area are likely to remain relevant even as the frontier of network design evolves.

This work is motivated by the perspective that everything in artificial intelligence is, at its heart, a matrix problem. Most machine learning and scientific workloads can be

A. Sabot, *Minimizing Data Movement and Parameter Count Across the Machine Learning Stack*, Synthesis Lectures on Computer Science,
https://doi.org/10.1007/978-3-032-23100-0_1

expressed as sequences of matrix and tensor operations, providing a hardware-agnostic interface for optimization across models and systems. Echoing Matt Levine's running joke from "Money Stuff"[1] that "everything in finance is securities fraud," this book argues that in artificial intelligence, all roads lead to matrices. Meaningful optimization eventually becomes a matter of efficiently representing and manipulating matrices and, increasingly, tensors (i.e., multi-dimensional matrices).

By focusing on matrices, one can expose and address fundamental bottlenecks such as memory bandwidth, communication, data reuse, and parallelism. As a result, this book adopts a matrix-centric viewpoint, in which AI workloads are understood and addressed primarily through the lens of matrix and tensor computations. The matrix-centric viewpoint is used to develop techniques that are robust to changes in model architecture, data type, and hardware, thereby demonstrating both the foundational and practical universality of the matrix as a focal point for AI system optimization.

An important contribution of this work is to enable users to run more complex models on smaller, resource-constrained user devices. When computation can occur locally, users gain the ability to keep their data private and secure, minimizing reliance on remote servers and reducing the risk of data exposure. This capability not only gives users more efficient and flexible options, but also helps close the gap in AI accessibility, supporting societal goals of autonomy, privacy, and fairness.

In the work presented here, we develop a collection of techniques including analytical tiling, hardware-compatible sparsity, low-rank matrix factorization, and distributed split computing that all use the lens of optimizing matrix and tensor computations for efficient machine learning. Rather than isolated solutions, these methods address complementary aspects of the same central challenge: optimizing the movement, storage, and transformation of data represented as matrices or tensors. Analytical tiling maximizes arithmetic intensity and hardware utilization, laying the groundwork for efficient dense computation. Sparsity and low-rank factorization, in turn, further reduce computational and memory demands by exploiting redundancy and structure in model parameters. Finally, distributed and split computing allow for flexible partitioning of machine learning workflows, enabling efficient use of both local and remote resources while supporting data privacy and adaptability.

By coordinating these approaches within a matrix-centric paradigm, my work offers a general and practical solution path for a wide range of machine learning problems, regardless of specific model architectures. Together, these techniques enable the deployment and training of sophisticated models on resource-constrained devices, the acceleration of large-scale systems, and the preservation of privacy in distributed environments.

[1] *Money Stuff* [3] is Matt Levine's popular Bloomberg Opinion newsletter, known for its witty and insightful commentary on finance, including the recurring observation that "everything in finance is securities fraud."

This book offers a systematic guide through these innovations, beginning each chapter with an accessible overview and proceeding to a more rigorous technical exposition and empirical results. The work aims to provide both fresh insight and practical frameworks that will be valuable to researchers, engineers, and practitioners seeking to deploy efficient, secure, and adaptive AI, regardless of how models and hardware continue to evolve.

1.2 Book Roadmap

The body of this book is organized as follows:

- **(Chapter 2) CAKE: Memory-Aware Block Shaping for High-Performance Blocked Matrix Multiplication** introduces an innovative algorithm for blocked matrix multiplication that optimizes block shapes and scheduling in order to match a target memory bandwidth, balancing memory access and compute to maximize throughput on modern hardware.
- **(Chapter 3) Multi-CAKE: Extending CAKE to Higher-Order Tensors** extends the bandwidth-aware block shaping principles introduced in CAKE from matrices to higher-order tensors. This chapter presents mCAKE, a framework for optimizing the memory and compute efficiency of dense tensor operations commonly found in deep learning and scientific computing. By systematically tiling and scheduling tensor computations, mCAKE adapts arithmetic intensity to satisfy memory bandwidth and size constraints across arbitrary tensor dimensions. When memory bandwidth is small, we need to increase arithmetic intensity. mCAKE demonstrates decreased bandwidth usage and speedups on modern hardware architectures.
- **(Chapter 4) Rosko: Sparse Matrix Multiplication for Machine Learning Workloads** describes Rosko (Row Skipping Outer Products for Sparse Matrix Multiplication Kernels), an approach for efficient matrix multiplication in the presence of moderate sparsity commonly found in deep neural networks. The chapter demonstrates how leveraging outer-products to skip zero computations can deliver gains on contemporary deep learning workloads. In addition, careful pruning of insignificant weights can induce efficient patterns of sparsity and further enhance performance.
- **(Chapter 5) Efficient Neural Network Computation with Rank-Sliced Gather-Scatter Activation** explores how neural network weights compressed via singular value decomposition (SVD) can be directly used for efficient inference. The method saves memory and computation through a gather-scatter approach operating on SVD components. Additionally, the gather-scatter approach is able to directly leverage sparsity in the SVD components to skip entire vector multiplication and additions.
- **(Chapter 6) Two-Stage Pruning and Sparsity-Preserving Fine-Tuning for SVD-Based Networks** details a two-stage framework for training SVD-compressed networks: first by truncating to a low-rank approximation and fine-tuning, then further

sparsifying and performing sparsity-preserving adaptation, thereby maintaining both efficiency and accuracy.

- **(Chapter 7) Block-by-Block Knowledge Distillation for Compressed Student Networks** introduces a block-wise knowledge distillation paradigm, where low-rank blocks of a student model are supervised directly by corresponding blocks of a full-rank teacher network, enabling efficient and performant training of highly compressed models.
- **(Chapter 8) Leveraging Cloud Knowledge Locally: Lightweight Edge-Side Classification with Cloud-Extracted Features** presents a split-computing framework empowering resource-constrained edge devices to benefit from powerful cloud-based extractors while maintaining user privacy. The chapter demonstrates rapid and customizable local classification using cloud-provided features, anchors, and user-side personalization.
- **(Chapter 9) Conclusion** concludes this book and contains final remarks.

References

1. Alexey Dosovitskiy, Lucas Beyer, Alexander Kolesnikov, Dirk Weissenborn, Xiaohua Zhai, Thomas Unterthiner, Mostafa Dehghani, Matthias Minderer, Georg Heigold, Sylvain Gelly, Jakob Uszkoreit, and Neil Houlsby. An image is worth 16 × 16 words: Transformers for image recognition at scale, 2021.
2. Andrei Ivanov, Nikoli Dryden, Tal Ben-Nun, Shigang Li, and Torsten Hoefler. Data movement is all you need: A case study on optimizing transformers, 2020.
3. Matt Levine. Money stuff. https://www.bloomberg.com/opinion/topics/money-stuff. Accessed: 2024-06-02.
4. Ashish Vaswani, Noam Shazeer, Niki Parmar, Jakob Uszkoreit, Llion Jones, Aidan N. Gomez, Lukasz Kaiser, and Illia Polosukhin. Attention is all you need, 2017.

2 CAKE: Memory-Aware Block Shaping for High-Performance Blocked Matrix Multiplication

Chapter Overview

Why the name CAKE? Most matrix multiplication algorithms view computation as a rigid, static grid. In CAKE, we view it as a dynamic resource-sharing problem, analogous to a host serving a cake to many guests. To ensure the party runs smoothly, the host must slice and serve portions at the exact rate the guests can consume them. If the host is too slow, the guests (compute cores) sit idle; if the host is too fast, the table (local memory) becomes cluttered and overwhelmed.

This analogy addresses a critical reality in modern AI: in demanding workloads like transformers, matrix multiplication is not limited by raw arithmetic throughput, but by the ability of the memory system to keep up. As matrix sizes scale, traditional methods make inefficient use of available memory bandwidth, leading to performance bottlenecks where the "movement" of data costs more than the computation itself.

By "shaping" our computational slices analytically, CAKE ensures that each core is optimally served a portion sized specifically to its memory and bandwidth constraints. This eliminates idle time and prevents memory congestion. This chapter moves from this intuitive "cake-cutting" philosophy to a rigorous mathematical framework designed to minimize data movement across the machine learning stack.

2.1 Introduction and Motivation

Over the last two decades, the explosion of machine learning workloads has shifted the primary challenge of high-performance computing from raw processing power to data management. While specialized hardware, such as Google's Tensor Processing Unit (TPU) [12], has emerged to meet this demand, the "memory wall" [25] remains a fundamental

A. Sabot, *Minimizing Data Movement and Parameter Count Across the Machine Learning Stack*, Synthesis Lectures on Computer Science,
https://doi.org/10.1007/978-3-032-23100-0_2

bottleneck. Addressing this disparity from a hardware perspective, e.g., by expanding on-chip SRAM, is often cost-prohibitive and physically constrained. Instead we focus on the software, we systematically maximize data reuse in the local memory to alleviate the disparity between memory and compute performance. In contrast, traditional approaches find optimal computation schedules through exhaustive grid searches of the parameter space, a process that becomes computationally intractable as systems scale.

Matrix-matrix multiplication (MM) serves as the computational backbone for many workloads, particularly in deep learning and scientific computing. In a standard convolutional neural network (CNN), for instance, the forward pass consists primarily of a sequence of matrix multiplications between layer inputs and weights [20]. The effective throughput of these operations is a multi-variable optimization problem involving processing power, internal bandwidth, local memory capacity, and external DRAM bandwidth. As processing power continues to increase, DRAM bandwidth frequently becomes the primary limiting factor, throttling overall performance despite available compute resources.

Current state-of-the-art approaches to matrix multiplication, most notably Goto's algorithm [7], are typically optimized for conventional systems where memory bandwidth and computational throughput are assumed to be balanced for standard workloads. However, the rapid diversification of hardware requires a more adaptive paradigm capable of navigating heterogeneous architectural characteristics. Such variations are increasingly common due to the emergence of special-purpose accelerators [6, 17], resource-constrained low-power systems [15, 18], 3D DRAM die-stacking technologies [3, 9, 13, 19], and high-capacity non-volatile memory (NVM) [22]. These evolving technologies create unique memory-to-compute ratios that necessitate a scheduling approach focused on analytical flexibility rather than fixed architectural assumptions.

Unlike fixed-library kernels that are hard-coded for the specific latency-bandwidth profiles of standard DDR4/5 memory, CAKE's analytical model treats the bandwidth-to-compute (BW:CT) ratio as a primary input variable. This mathematical flexibility allows the framework to automatically adapt its block geometry for the massive throughput of 3D die-stacked DRAM [13] or the high-latency, asymmetric read/write characteristics of Non-Volatile Memory [22] without requiring manual re-tuning or exhaustive empirical searches. This adaptability remains critical even with the next generation of memory technologies.[1] Despite these generational leaps, the fundamental imbalance persists; such hardware advancements effectively feed the "memory wall" rather than dismantling it, as the computational requirements of trillion-parameter models continue to outpace the growth of memory throughput [21].

This chapter addresses the fundamental imbalance between computational power and memory bandwidth through an analytical scheduling framework. We demonstrate

[1] For example, DDR6 aims to double the base bandwidth over its predecessor [11] and the commercial rollout of HBM4 will have a 2048-bit interface [14].

that CAKE can significantly improve matrix multiplication throughput across diverse architectures, including Intel, AMD, and ARM processors. Following an overview of matrix multiplication and the foundational block framework in Sect. 2.3, we detail the two primary scheduling techniques introduced in CAKE [16] to outperform state-of-the-art libraries on fixed hardware.

The first technique, discussed in Sect. 2.4, involves analytical block shaping and sizing. While prior methods such as autoblock [23] increase arithmetic intensity by expanding a single block dimension, they fail to account for the constraints of external memory bandwidth. In contrast, CAKE precisely regulates arithmetic intensity by matching the computation time of local memory blocks with the latency of external memory accesses.

The second technique focuses on optimizing the traversal order of block computations to maximize inter-block data reuse within local memory (Sect. 2.5). While traditional compilers [2,4] typically optimize loop nests by minimizing the iteration distance between redundant accesses, CAKE further reduces external DRAM traffic by integrating block geometry directly into the scheduling logic. Finally, Sect. 2.6 provides a comprehensive performance evaluation of these techniques on a modern desktop processor.

2.2 Why the Name CAKE?

In CAKE, we partition the matrix multiplication computation space—a 3D volume of multiply-accumulate (MAC) operations—into a grid of computation blocks. These blocks are scheduled and then sequentially executed across multiple compute cores (Fig. 2.1).

This scheme is analogous to how a host cuts and serves a "cake" to their guests. Here, the size of the cake represents the total computational workload to be performed. The surfaces of each slice correspond to the amount of data exchanged between different levels of memory: the input/output (IO) required to move data in and out of local storage. Each guest is given a plate, which represents the limited local memory available to each compute core.

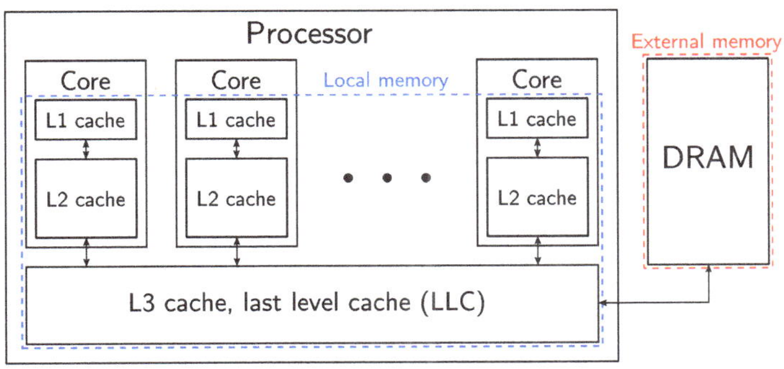

Fig. 2.1 Computing architecture with many cores, multiple levels of local memory, and external memory

To consume the entire cake as quickly as possible, each guest must receive a steady stream of cake without waiting. In our analogy, the rate at which the host serves cake must match the guests' rate of consumption, just as in CAKE the scheduling and division of computational blocks is orchestrated to keep all cores busy and maximize throughput, minimizing idle time due to waiting for data.

This cake-cutting analogy captures the goals and principles of the CAKE algorithm: carefully partitioning and distributing work such that every processing element is optimally served, making the most efficient use of local memory and minimizing data transfers.

2.3 Background on Matrix Multiplication

Consider a matrix multiplication between matrices A and B, where A is size $M \times K$ and B is size $K \times N$. The matrix multiplication can be computed via a set of vector operations using one of the two strategies. That is, we can obtain the matrix multiplication result C through $M \cdot N$ inner products between M row vectors (size $1 \times K$) of A and N column vectors (size $K \times 1$) of B, or summation of K outer products between column vectors (size $M \times 1$) of A and the corresponding row vectors (size $1 \times N$) of B. Algorithm 1 defines matrix multiplication where reduction occurs on dimension K.

Outer products (Fig. 2.2a), unlike inner products (Fig. 2.2b), yield partial result matrices which will be summed together to produce C, allowing reuse and in-place accumulation. In the $A \times B$ example (Fig. 2.3), there are K outer products between vectors of size $M \times 1$ and $1 \times N$, which each produce partial result matrices of size $M \times N$. Partial results are accumulated across the K dimension (reduction dimension). Note that we may store the intermediate partial result matrix locally to be accumulated in place with forthcoming partial result matrices. Thus, the locally stored intermediate results are reused K times.

Algorithm 1: Matrix multiplication (outer product strategy)

```
for i = 1 → M do
  for j = 1 → N do
    for k = 1 → K do
      C[i][j] ⟵ C[i][j] + A[i][k] · B[k][j];
```

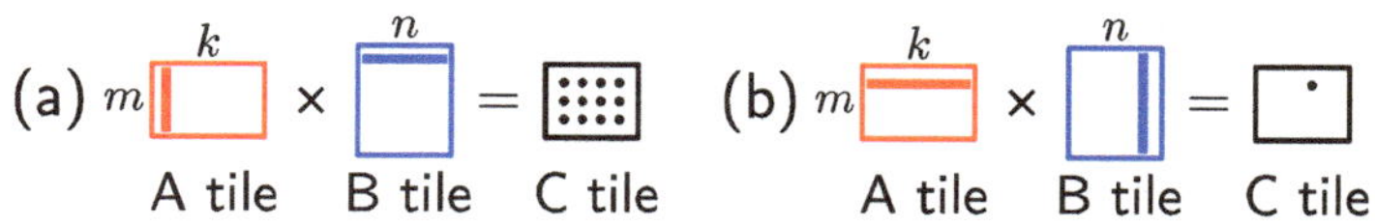

Fig. 2.2 (**a**) Outer product and (**b**) inner product tile MM. The dots in C represent elements computable with the highlighted columns and rows of the red A and blue B tiles. Outer products compute partial results for a C tile using a column vector from the A tile and a corresponding row vector from the B tile. Inner products fully accumulate a single C element using a row vector from the A tile and a column vector from the B tile

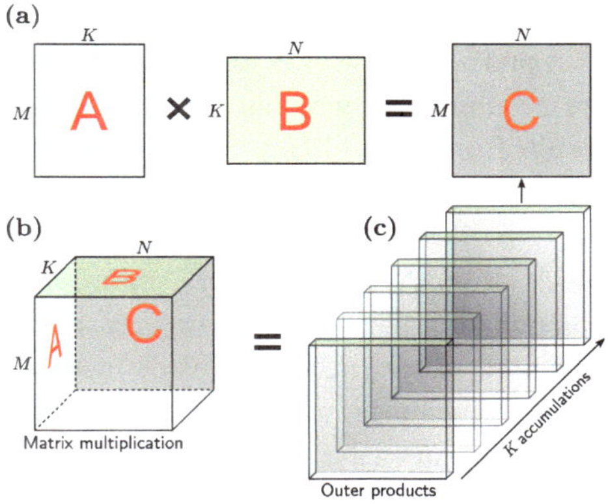

Fig. 2.3 (**a**) $C = A \times B$ matrix multiplication. (**b**) Computation space represented as an $M \times K \times N$ 3D volume of multiply-accumulate (MAC) operations as defined by Algorithm 1. (**c**) Computation space as an accumulation of outer products

The **computation space** for $C = A \times B$ is a 3D volume of $M \times N \times K$ multiply-accumulate (MAC) operations depicted in Fig. 2.3b. The volume shown in Fig. 2.3b is determined by 3 I/O surfaces: we consider matrix A as being on the "left" side, matrix B being on the "top", and matrix C being the "back wall".

2.3.1 Block Framework for MM Computation

If a computing platform has a sufficient local memory (internal memory) to hold the 3 I/O surfaces, all MACs in the 3D computation space may be completed without any additional I/O operations. For large matrix computations exceeding local memory capacity, the computation space must be partitioned and computed in smaller blocks, whose results are later combined.

As shown in Fig. 2.5, we partition the $M \times N \times K$ computation space into smaller blocks of $m \times n \times k$ elements. Section 2.4 describes how block dimensions are calculated based on available resources, such as total available computing power and external DRAM bandwidth. Computation of a block may be viewed as a sum of k outer products, for multiplying the two corresponding sub-matrices within the larger matrix multiplication computation space. The computation yields the back wall of the block. Block computations result in $m \times n$ sub-matrices which, when accumulated together, yield a portion of the final matrix C. Matrix addition is commutative, so computation order for blocks in the computation space does not matter for correctness.

All cores in the processing grid (depicted in Fig. 2.5b) work in parallel, on input tiles at the rate of one tile result per unit time for each core, to compute a block by performing k outer products. Each core works through the N-dimension of the block computation space, producing a row of partial results by computing tile-wise multiplications between a single A tile and n B tiles. It is also possible to compute computation blocks in the M or K-dimension but we focus our presentation on the N-dimension. Each column of cores in the grid (e.g., cores 1, 5, 9, 13 in Fig. 2.5b) computes an outer product between a sub-column of A and a sub-row of B to produce a partial result sub-matrix. The resulting sub-matrices are accumulated, yielding a partial sub-matrix which will be further accumulated with results from other blocks.

Intra-block data movement is reduced by each core sequentially reusing one A tile with many B tiles. The B tiles are broadcast to cores in the same column to maximize intra-block reuse. Partial results are summed along the K dimension (towards the back of the computation space), maximizing reuse via in-place accumulation.

As noted earlier, a block is defined by three I/O surfaces: an input surface A of size $m \times k$, an input surface B of size $k \times n$, and a result surface C of size $m \times n$. Surfaces A and B correspond to the aforementioned sub-matrices within the larger matrix multiplication computation space. Depending on the location of the block within the computation space, result surface C will consist of either partial or completed reduction results.

For each block, the total external memory I/O and required local memory size are both equal to the sum of the three I/O surfaces. External memory bandwidth requirements may be determined from the computation time of the block. When the block is shaped properly (see Fig. 2.4), the I/O time for the three surfaces will match the computation time of the block, allowing I/O to overlap computation that fully utilizes available processing power.

I/O requirements are further reduced when sequentially computed blocks share an I/O surface (i.e., the blocks are adjacent within the computation space). The surface can be kept local (stationary): the following adjacent block can *reuse* the surface without needing to fetch it from external memory. We assume that each C-tile requires two I/O accesses, one to read it and another to write it back to external memory. Furthermore, if the surface is a partial result surface, the previous block does not need to write back the results to external memory before the next computation. I/O surfaces can be shared in the M, N, or K-dimension, and I/O cost is minimized when the largest I/O surface is reused most frequently.

2.4 Computation Block Shaping

A constant bandwidth (CB) block is a block (described in Sect. 2.3.1) with dimensions (n, m, k) shaped and sized according to the external bandwidth and available compute resources (as seen in Fig. 2.4). CB block shaping provides control over **arithmetic intensity** (AI), allowing us to match external I/O time with computation time. AI is defined as the ratio of computation volume to data transferred, which is equivalent to the ratio of computation throughput (CT) to external memory bandwidth (BW): $AI = \frac{CT}{BW}$. Therefore, we can, for example, increase CT or decrease required BW by using CB blocks to control AI. We can also increase the size of the CB block to leverage any additional BW and decrease internal memory size requirements.

Consider a computing architecture with a number of cores, each performing one tile multiplication per unit time. As seen in Sect. 2.3.1, each core handles one tile from A, so

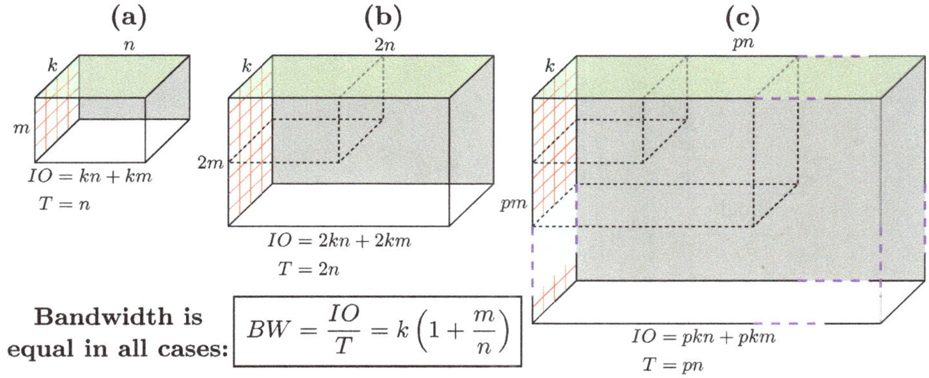

Fig. 2.4 Changing the shape and size of a block can keep a block's external bandwidth (BW) constant when increasing the computation throughput (CT) by increasing core count. The block gets taller and wider, matching IO and computation time T as volume (V) changes. We only factor in the I/O of the A and B surfaces as we assume the 3rd surface C is kept stationary. Note that *arithmetic intensity* is V/IO. Since $V/IO = \frac{V/T}{IO/T} = \frac{CT}{BW}$, and constant bandwidth blocks in (**a**), (**b**) and (**c**) have equal BW and increasing volume, they have increasing arithmetic intensity. Importantly, **computation throughput** is V/T. Thus, CB blocks in (**a**), (**b**) and (**c**) have increasing computation throughput at $\frac{mkn}{n} = mk$, $\frac{4mkn}{2n} = 2mk$, and $\frac{p^2mkn}{pn} = pmk$, respectively

the number of tiles in the A surface (size $m \times k$) of a CB block is equal to the number of cores. The size of k determines how many cores contribute their partial results for accumulation. In this analysis, we assume k is a fixed optimal value calculated from available external bandwidth (Sect. 2.4.2). Given k, we can compute m so that $m \cdot k$ is the number of available cores (Fig. 2.5a, b). We reflect an increase in the amount of cores used in p (e.g., when trying to increase CT by doubling the number of cores used, p would increase by a factor of 2). To reduce the number of variables in our analysis, we set m to a multiple of k, based on the number of available cores (i.e., $m = pk$), but m does not need to be a multiple of k.

Thus, the CB block shape is defined by $m = pk$ and $n = \alpha pk$ where $\alpha \geq 1$ and k are unitless constants calculated from available external memory bandwidth (see Sect. 2.4.2). When external memory bandwidth is low, raising α increases block computation time, thereby decreasing the CB block's external bandwidth requirement (BW). When there is sufficient external bandwidth, α is set to 1.

To compute a CB block in the N-dimension, each core is first loaded with one A tile. B tiles are then streamed to each core from local memory (e.g., L3 cache). The CB block is shaped to have exactly one A tile per core, keeping A tiles stationary, to reduce local congestion. Results are accumulated between cores and cycled back to local memory (e.g., L3 cache) for reuse and moved to external DRAM when complete. The computation time T for a CB block is $n = \alpha pk$ unit times because each core is assigned to compute n tile multiplications.

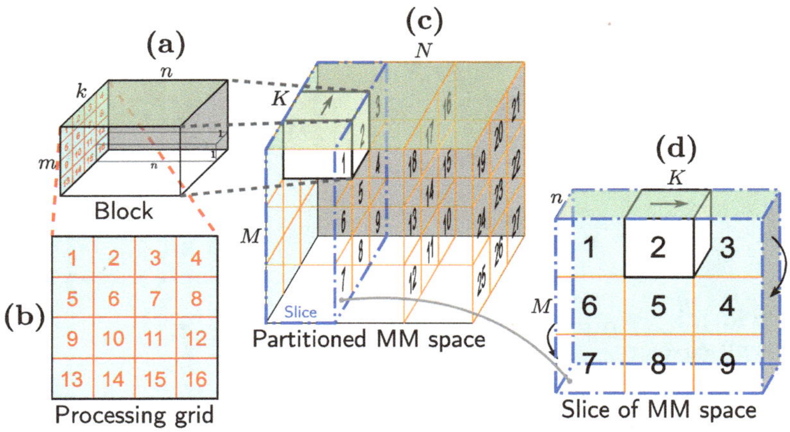

Fig. 2.5 (**a**) Block defined by input and output surfaces: a blue A input tile ($k \times m$), a green B input tile ($n \times k$), and a gray C output tile ($m \times n$). (**b**) Grid of 16 processing cores. Red numbers represent individual cores. (**c**) Block-partitioned computation space for an MM between $M \times K$ and $K \times N$ matrices. (**d**) Rotated view of a slice of the computation space. The black numbers represent the order of execution for blocks in a K-first schedule

Alternatively, we can compute a CB block in the M or K-dimension, resulting in a CB block computation time of k or m unit times, respectively. Computing CB blocks in alternative directions may be advantageous on certain architectures. For example, computing CB blocks in the K-dimensions is preferable when doing in-place accumulation. In this analysis, we do not factor in accumulation time as we assume accumulation can be overlapped with multiplication.

2.4.1 Internal Memory Size Requirement

A CB block consists of 3 I/O surfaces that must be stored locally. The I/O for each surface is equal to its size. For A, $IO_A = pk \cdot k = pk^2$. For B, $IO_B = k \cdot \alpha pk = \alpha pk^2$. For result surface C, $IO_C = pk \cdot \alpha pk = \alpha p^2 k^2$. The internal memory size requirement is simply the combined size of the surfaces:

$$MEM_{internal} = IO_A + IO_B + IO_{partial} = \alpha pk^2 + pk^2 + \alpha p^2 k^2$$

To increase the target processing power p-fold, internal memory size must increase by a factor of p^2 (due to the third term, $\alpha p^2 k^2$).

2.4.2 External Bandwidth Analysis

We may compute the minimum external bandwidth required for a CB block based on the external I/O for its A and B surfaces, using the following equation, where T is the block computation time:

$$BW_{min} = \frac{IO_A + IO_B}{T} = \frac{\alpha pk^2 + pk^2}{\alpha pk} = \left(\frac{\alpha + 1}{\alpha}\right) \cdot k \text{ tiles/cycle}$$

Increasing α allows us to compensate for low external bandwidth but increases both computation time and required local memory size. Partial result I/O is not considered since results are held locally.

We define external bandwidth as $BW_{ext} = R \cdot k$ tiles/cycle, where $R > 1$ is a constant capturing the difference between available external bandwidth and the minimum bandwidth defined previously. We satisfy the minimum external bandwidth requirement when $BW_{ext} \geq BW_{min}$, or $\alpha \geq \frac{1}{R-1}$.

Now consider the case when more processing power is available (e.g., when the number of cores increases from 16 to 32, as in Fig. 2.4). We choose to increase the N and M-dimensions by a factor of $p = 2$ because I/O and computation time will increase by the same factor. BW_{min} does not depend on p, which increases both computation and I/O time equally. As a result, we can increase the number of utilized cores (pk^2) while keeping the same external bandwidth requirement (see Fig. 2.4).

2.4.3 Internal Bandwidth Requirements

Recall that the local memory holds 3 I/O surfaces: two input surfaces and one result surface. During a CB block computation, each input surface is read once and loaded onto the cores. The partial result surface is accessed twice: once for reading and once for storing new partial results. We see the internal bandwidth must be at least:

$$\frac{IO_A + IO_B + 2IO_{partial}}{T} = Rk + 2pk \text{ tiles/cycle}$$

Thus, as the number of cores (pk^2) increases, internal bandwidth must increase proportionally (due to the second term, $2pk$) to match external I/O time with computation time.

2.5 Computation Block Scheduling

Along with shaping the CB block we must also schedule the order of block computations such that adjacent blocks are computed in sequence which minimizes external memory accesses. Using both the CB block shape and size and the matrix multiplication operand dimensions we can pick a schedule which minimizes external memory accesses.

2.5.1 Choosing Loop Ordering Based on I/O (Memory Accesses)

To illustrate how loop order impacts external memory accesses, we demonstrate the difference in total external memory accesses for two loop orders of the same matrix multiplication.

Referring to the partitioned multiplication between an $M \times K$ matrix A and a $K \times N$ matrix B described in Fig. 2.5, the total I/O for a given loop order can be computed as a sum of the streaming (data fetched between each block) and stationary I/O (data reused between blocks):

$$(\text{\# of blocks}) \cdot (IO_{streaming} \text{ per block}) + IO_{stationary}$$

For an N-first schedule (N as the inner loop), our total I/O is:

$$\begin{aligned} IO_{N\text{-first}} &= \frac{M}{m} \cdot \frac{K}{k} \cdot \frac{N}{n}(mn + nk) + \frac{M}{m} \cdot \frac{K}{k}(mk) \\ &= MKN\left(\frac{1}{m} + \frac{1}{k}\right) + MK \end{aligned}$$

For an M-first schedule, our total I/O is:

$$\begin{aligned} IO_{M\text{-first}} &= \frac{M}{m} \cdot \frac{K}{k} \cdot \frac{N}{n}(mk + mn) + \frac{K}{k} \cdot \frac{N}{n}(kn) \\ &= MKN\left(\frac{1}{k} + \frac{1}{n}\right) + KN \end{aligned}$$

Consider a matrix multiplication between A and B where $M > N, K$ computed with square tiles ($n = m = k$). Computing the matrix multiplication in the N-dimension first requires more total external memory accesses than if we computed in the M-dimension first. In comparison, the M-first order has less I/O since B tiles can be reused $\frac{M}{m}$ times, which is more than other directions because $M > N, K$. The streaming term of the I/O function is independent of loop order when tile sizes are equal. The stationary term of the I/O function is dependent on loop order because the matrix multiplication size determines stationary I/O. Figure 2.6 shows how choice of M-first or K-first schedules impacts computation throughput for matrices where M or K is large.

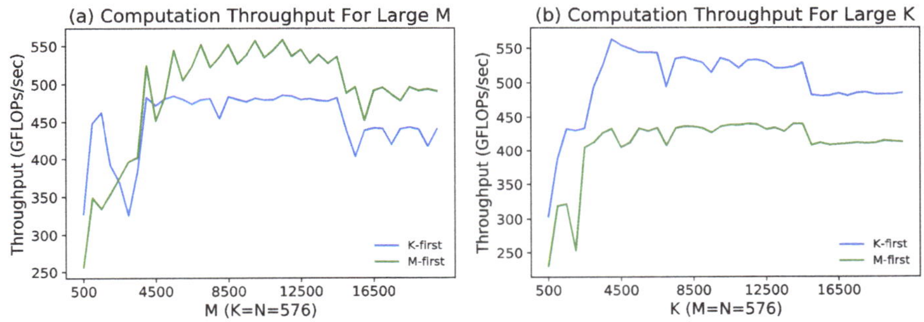

Fig. 2.6 Computation throughput on an Intel i9-10900k for different schedules and matrix shapes. We fix two dimensions at 576 and vary the third (M or K) from 500 to 20,000. (**a**) Shows that, when M is larger than N and K, the M-first schedule minimizes I/O and achieves higher throughput than the K-first schedule. (**b**) Shows the K-first schedule is optimal when K is larger than N and M

2.5.2 Choosing Loop Ordering Based on Tile Dimensions Minimizes External Memory Accesses

Consider a matrix multiplication where $M = K = N$ with non-square tile sizes: $n > k = m$. If we use the optimized loops for a square matrix given in [2], we obtain an N-first schedule. The above N-first method does not account for non-square tile sizes and is therefore streaming more data than necessary. In an N-first loop order, an A tile is kept stationary while B and C tiles are streamed. B and C tiles are larger and thus require more I/O to stream. We can reduce I/O by going M-first or K-first and streaming A tiles. For this example, M-first and K-first are equivalent since all other dimensions are equal. To minimize external memory accesses, tile size should be determined *first* instead of tiling *after* determining loop order.

After determining a tile size, there is sufficient information to calculate both the streaming and stationary I/O costs of any loop order. Tile size determines the number of memory accesses associated with streaming and the size of A and B determines the memory accesses associated with the stationary data. A loop order that minimizes external memory accesses may be selected by calculating the memory accesses for each of the 6 possible loop orders of M, K, N.

2.5.3 Reuse Between Dimensions

To allow inter-block reuse between different dimensions (e.g., between K and M), the computation space traversal must "turn" every time it completes a dimension. Consider the loop order: N, M, K. If the loops always started at the 0 index of a dimension, no A or B surfaces would be reused, leading to $O(MN+N)$ missed I/O surface reuses. The traversal direction flips after each dimension to allow for I/O surface reuse, shown in Algorithm 2.

Algorithm 2: K-first block partitioning algorithm

```
// Get the number of blocks in each dimension
M_b = M/m; N_b = N/n; K_b = K/k
for n_idx = 0, 1, ... to N_b − 1 do
    // Flip direction of M traversal for A reuse
    if n_idx mod 2 == 0 then m_start = 0; m_end = M_b
    else m_start = M_b; m_end = 0
    for m_idx = m_start to m_end do
        // Flip direction of k traversal for B reuse
        if n_idx mod 2 == 0 then
            if m_idx mod 2 == 0 then k_start = 0; k_end = K_b
            else k_start = K_b; k_end = 0;
        else
            if m_idx mod 2 == 0 then k_start = K_b; k_end = 0;
            else k_start = 0; k_end = K_b;
        for k_idx = k_start to k_end do
            // Compute inner block multiplication
            C[m_idx * m][n_idx * n]  += A[m_idx * m][k_idx * k] × B[k_idx * k][n_idx * n];
```

Note the pseudocode assumes $N \geq M$, when $M > N$ the outer two loops are switched because the A surfaces should be reused before the B surfaces. The algorithm defines the K-first computation order of blocks, which sweeps the space of computation space by first traversing the K-dimension to maximize partial result reuse, then the M-dimension to reuse A, and lastly the N-dimension to reuse B.

2.6 Evaluation

This section showcases results from [16], where the techniques described in Sects. 2.4 and 2.5 are combined. Current state-of-the-art matrix multiplication algorithms are based on GOTO's algorithm [8] and are bounded by external memory bandwidth.

We implemented CAKE in C++ for use in our evaluations. Our implementation uses the BLIS kernel library [24] to execute matrix multiplication at the tile level on CPU SIMD registers. BLIS was chosen for its portability across CPU architectures, enabling CAKE to act as a drop-in GEMM library on multiple platforms.

We compare CAKE to OpenBLAS when computing a matrix multiplication between two large 23040×23040 matrices on an AMD Ryzen 5950X CPU. OpenBLAS was chosen for the AMD CPU because it is an optimized library using GotoBLAS and outperforms MKL [1] on non-Intel hardware [5]. We use AMD μProf [10] to measure DRAM bandwidth and computation throughput. However, due to the lack of an available DRAM access counter on this CPU, DRAM accesses during matrix multiplication are estimated using the PMU counter for L1 data cache refills.

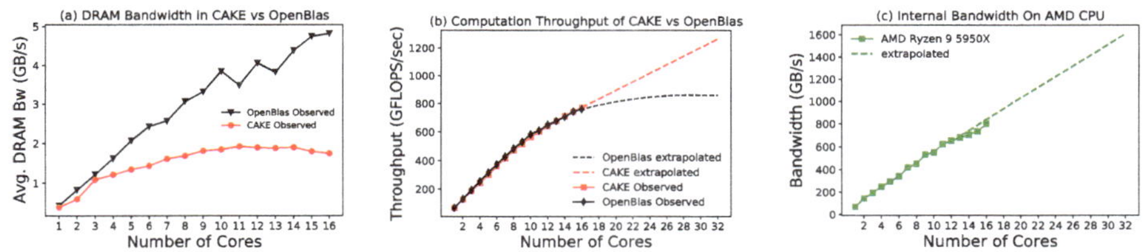

Fig. 2.7 MM between two 23,040 × 23,040 matrices on an AMD Ryzen 9 5950X CPU using CAKE and OpenBLAS. (**a**) Unlike OpenBLAS (black), CAKE (red) does not need to increase DRAM bandwidth to utilize more cores. On this CPU, DRAM bandwidth is estimated from the number of L1 cache-line refills from DRAM. (**b**) CAKE matches OpenBLAS's peak performance, but with a smaller DRAM bandwidth requirement, as seen previously. (**c**) Using *pmbw*, we measured internal bandwidth between the L3 cache and cores. The 5950X has high internal bandwidth between L3 cache and cores. This bandwidth grows roughly linearly with the number of cores. Consequently, we see in (**a**) that CAKE's required DRAM bandwidth stays constant past 9 cores, and in (**b**) that sufficient internal bandwidth enables CAKE to achieve high computation throughput. The dotted extrapolation lines assume internal memory bandwidth increases proportionally for each additional core, local memory size increases quadratically, and DRAM bandwidth is fixed

Figure 2.7c shows internal bandwidth between the L3 cache and processor cores increases roughly proportionally by $50\,\mathrm{G\,B\,s^{-1}}$ per core. CAKE takes advantage of the increasing internal bandwidth to achieve peak computation throughput without increasing DRAM bandwidth usage beyond 9 cores (Fig. 2.7a, b). Since the CPU's internal bandwidth and DRAM bandwidth are sufficient, both CAKE and OpenBLAS can increase computation throughput when adding more cores, but OpenBLAS uses more DRAM bandwidth than CAKE.

We include extrapolations for expected performance for CAKE and OpenBLAS [26], again assuming internal memory bandwidth continues to increase proportionally with the number of cores while DRAM bandwidth remains fixed in Fig. 2.7b, c.

2.7 Summary

The work presented in this chapter demonstrates that while the "memory wall" remains a physical constraint of modern hardware, its impact on performance can be significantly alleviated through intelligent, bandwidth-aware scheduling. By replacing computationally expensive grid searches with analytically-derived Constant-Bandwidth (CB) blocks, CAKE provides a deterministic method to maximize data reuse and match computation time with memory latency.

In the current landscape of DDR6 and HBM4, where the gap between compute potential and data delivery continues to widen, this approach serves as a critical bridge. CAKE's ability to maintain near-constant DRAM bandwidth usage while scaling core counts proves that we can extract higher utility from existing memory hierarchies. By treating the bandwidth-to-compute ratio as a first-class variable, we move from a paradigm of hardware-specific tuning to one of analytical flexibility. This foundation ensures that as models scale toward trillion-parameter horizons, the movement of data remains a manageable and optimized component of the machine learning stack. The following chapter (Chap. 3) will explore how these principles of blocked matrix multiplication are extensible to a broader class of higher-order tensor contractions.

References

1. *Intel oneAPI Math Kernel Library*.
2. Nawaaz Ahmed, Nikolay Mateev, and Keshav Pingali. Synthesizing transformations for locality enhancement of imperfectly-nested loop nests. In *Proceedings of the 14th International Conference on Supercomputing*, ICS '00, page 141–152, New York, NY, USA, 2000. Association for Computing Machinery.
3. B. Black, M. Annavaram, N. Brekelbaum, J. DeVale, L. Jiang, G. H. Loh, D. McCaule, P. Morrow, D. W. Nelson, D. Pantuso, P. Reed, J. Rupley, S. Shankar, J. Shen, and C. Webb. Die stacking (3D) microarchitecture. In *2006 39th Annual IEEE/ACM International Symposium on Microarchitecture (MICRO'06)*, pages 469–479, 2006.
4. Uday Bondhugula, Muthu Baskaran, Sriram Krishnamoorthy, Jagannathan Ramanujam, Atanas Rountev, and Ponnuswamy Sadayappan. Automatic transformations for communication-minimized parallelization and locality optimization in the polyhedral model. In *International Conference on Compiler Construction*, pages 132–146. Springer, 2008.
5. Agner Fog. Intel's 'cripple amd' function. https://www.agner.org/optimize/blog/read.php?i=49, 2020.
6. Mingyu Gao, Jing Pu, Xuan Yang, Mark Horowitz, and Christos Kozyrakis. Tetris: Scalable and efficient neural network acceleration with 3D memory. In *Proceedings of the Twenty-Second International Conference on Architectural Support for Programming Languages and Operating Systems*, ASPLOS '17, page 751–764, 2017.
7. Kazushige Goto and Robert A. van de Geijn. On reducing TLB misses in matrix multiplication. https://users.umiacs.umd.edu/~ramanid/cmsc662/Goto_vdGeijn.pdf, 2002.
8. Kazushige Goto and Robert A. van de Geijn. Anatomy of high-performance matrix multiplication. *ACM Trans. Math. Softw.*, 2008.
9. J.H. Huang. Keynote. NVidia GPU Technology Conference, 2013.
10. Advanced Micro Devices Inc. AMD μprof. https://developer.amd.com/amd-uprof/, 2021.
11. JEDEC Solid State Technology Association. JESD79-6: DDR6 SDRAM standard. 2025. Initial specification for doubling bandwidth over DDR5.
12. Norman P. Jouppi, Cliff Young, et al. In-datacenter performance analysis of a tensor processing unit. *CoRR*, abs/1704.04760, 2017.
13. H. Jun, J. Cho, K. Lee, H. Son, K. Kim, H. Jin, and K. Kim. HBM (high bandwidth memory) DRAM technology and architecture. In *2017 IEEE International Memory Workshop (IMW)*, pages 1–4, 2017.
14. J. Kim and S. Lee. Mass production of HBM4: Addressing the 2048-bit interface for trillion-parameter models. In *2026 IEEE International Solid-State Circuits Conference (ISSCC)*, 2026.
15. Ashish Kumar, Saurabh Goyal, and Manik Varma. Resource-efficient machine learning in 2 kb ram for the internet of things. In *Proceedings of the 34th International Conference on Machine Learning - Volume 70*, ICML'17, page 1935–1944. JMLR.org, 2017.

16. H. T. Kung, Vikas Natesh, and Andrew Sabot. CAKE: Matrix multiplication using constant-bandwidth blocks. In *Proceedings of the International Conference for High Performance Computing, Networking, Storage and Analysis*, SC '21. Association for Computing Machinery, 2021.
17. Z. Li, Z. Wang, L. Xu, Q. Dong, B. Liu, C. I. Su, W. T. Chu, G. Tsou, Y. D. Chih, T. Y. J. Chang, D. Sylvester, H. S. Kim, and D. Blaauw. RRAM-DNN: An RRAM and model-compression empowered all-weights-on-chip DNN accelerator. *IEEE Journal of Solid-State Circuits*, 56(4):1105–1115, 2021.
18. Ji Lin, Wei-Ming Chen, Yujun Lin, John Cohn, Chuang Gan, and Song Han. MCUNet: Tiny deep learning on IoT devices, 2020.
19. Jason Lowe-Power, Mark D. Hill, and David A. Wood. When to use 3D die-stacked memory for bandwidth-constrained big data workloads, 2016.
20. Bradley McDanel, Sai Qian Zhang, HT Kung, and Xin Dong. Full-stack optimization for accelerating CNNs using powers-of-two weights with FPGA validation. In *Proceedings of the ACM International Conference on Supercomputing*, pages 449–460, 2019.
21. Micron Technology. 2026 memory market outlook: The shift to AI-centric HBM production. Technical report, Micron Investor Relations, 2026.
22. Dennis Rich, Andrew Bartolo, Carlo Gilardo, Binh Le, Haitong Li, Rebecca Park, Robert Radway, Mohamed Sabry, H.-S. Philip Wong, and Subhasish Mitra. Heterogeneous 3D nano-systems: The n3xt approach?, 06 2020.
23. Robert Schreiber, Jack J Dongarra, et al. *Automatic blocking of nested loops*. Research Institute for Advanced Computer Science, NASA Ames Research Center, 1990.
24. Field G. Van Zee and Robert A. van de Geijn. BLIS: A framework for rapidly instantiating BLAS functionality. *ACM Transactions on Mathematical Software*, 41(3):14:1–14:33, June 2015.
25. Wm. A. Wulf and Sally A. McKee. Hitting the memory wall: Implications of the obvious. *SIGARCH Comput. Archit. News*, 23(1):20–24, March 1995.
26. Zhang Xianyi, Wang Qian, and Zaheer Chothia. Openblas. *URL:* http://xianyi.github.io/OpenBLAS, page 88, 2012.

Multi-CAKE: Extending CAKE to Higher-Order Tensors

3

Chapter Overview

In the previous chapter, we established that matrix multiplication, representing the classic three-loop problem, could be optimized through the analytical shaping of computational blocks. However, matrix multiplication is merely a specific case within a much broader landscape of n-dimensional tensor operations. As we transition from matrices to higher-order tensors, we encounter a combinatorial explosion where the number of possible loop permutations and tiling strategies grows exponentially with the depth of the nest.

This chapter introduces multi-CAKE (mCAKE), a framework designed to manage this complexity by generalizing the constant-bandwidth principle. The name reflects a fundamental shift in perspective. Rather than viewing a multi-dimensional contraction as a single, monolithic volume, we decompose the computation into a structured sequence of matrix-centric operations. This is represented conceptually as a set of multiple CAKEs, as illustrated in Fig. 3.3.

mCAKE addresses the rapidly expanding search space that arises as we move beyond three-dimensional matrix multiplication to more general multi-dimensional tensor operations, a common scenario in modern machine learning workloads, such as convolutional layers and transformer models. While this greater complexity might suggest the need for increasingly elaborate optimization strategies, mCAKE demonstrates that we can still achieve optimal bandwidth usage by simply applying the CAKE kernel to the innermost three loops. This not only streamlines the implementation but also provides valuable insight: the innermost loops dominate data reuse and, thus, bandwidth efficiency, regardless of the outer structure.

Throughout this chapter, we explore how mCAKE adapts the constant-bandwidth property of CAKE from matrices to higher-dimensional tensors, and describe its practical relevance through examples such as parallel matrix multiplications and generalized

A. Sabot, *Minimizing Data Movement and Parameter Count Across the Machine Learning Stack*, Synthesis Lectures on Computer Science,
https://doi.org/10.1007/978-3-032-23100-0_3

convolutions. By building on the foundation laid by CAKE, mCAKE offers a path to efficient, scalable computation for a wide range of advanced linear algebra operations.

3.1 Introduction

Matrix multiplication, covered in the previous chapter, is only the simplest member of a much richer family of high-dimensional tensor operations. In practice, *tensor contractions* (e.g. Einstein-summation) and direct *convolutions* dominate the arithmetic count of modern machine learning and scientific codes: forward passes of convolutional neural networks comprise dozens of 2-D convolutions per layer [22], and contractions account for more than 99% of the floating-point operations in some transformer workloads [15]. Achieving high performance for these kernels on multi-core CPUs is now primarily limited by *off-chip memory bandwidth*, not by raw computational throughput: a widening "memory wall" illustrated in Fig. 3.1.

Classical compiler optimizations, namely *tiling* (blocking) and *loop reordering*, reduce data movement by reusing data that fits in on-chip caches [3, 17, 23]. Unfortunately, the design space explodes: for an L-level memory hierarchy and an n-deep loop nest, the number of tile size and permutation choices grows exponentially. Therefore state-of-the-art approaches rely on heuristic search [5, 8], auto-tuning supported by analytical cost models [19, 20], or pruning via linear programming—all costly and still not guaranteed to minimize data movement.

From CAKE to Multi-CAKE The CAKE framework in Chap. 2 demonstrated that, for the *three-loop* matrix-multiplication kernel, one can derive a *constant-bandwidth* schedule analytically, removing the need for auto-tuning [16]. This chapter extends that philosophy to *arbitrary* n-deep loop nests (tensor contractions and $k \geq$ 3-D convolutions). Our method—**multi-CAKE** (mCAKE)—selects the three innermost loops with the highest

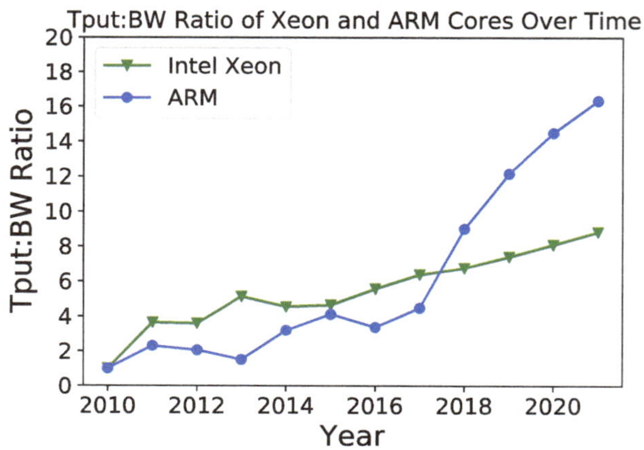

Fig. 3.1 Ratio of computation throughput (Tput) to off-chip DRAM memory bandwidth (BW) for server-class (Intel Xeon) and embedded (ARM Cortex-A) processor families over the years. Each data point represents a particular architecture version and was collected from [28, 29]

reuse potential, applies CAKE to them, and derives analytical tile sizes for the outer loops so that total DRAM traffic is provably minimal under the given cache capacity.

For convolutions whose stride is smaller than the kernel (stride $=$ 1, kernel 3×3, etc.), common "lowering" techniques replicate input patches and inflate memory traffic. mCAKE instead works directly on the native tensor layout, avoiding duplication (Sect. 3.5).

Why an Analytical Model Matters Modern CPUs provide ample arithmetic throughput and on-chip bandwidth; off-chip DRAM, however, improves slowly (Fig. 3.1). An analytical schedule that *guarantees* near minimal DRAM traffic allows performance to scale with added cores without increasing memory bandwidth. Keeping memory bandwidth low is critical for both speed and energy. This comes with the drawback of needing increased local memory.

Contributions of This Chapter

- We derive an analytic tiling model—mCAKE—that minimizes DRAM traffic for deep, multi-level loop nests on multi-core CPUs, entirely eliminating design-space exploration (Sect. 3.4).
- We extend CAKE's three-loop schedule to tensor contractions and direct convolutions, including stride-1 kernels without data replication (Sect. 3.5).
- We empirically demonstrate higher throughput and lower DRAM bandwidth than state-of-the-art auto-tuned and model-search methods on several Intel and ARM processors (Sects. 3.6.3 and 3.6.4).

The remainder of the chapter is organized as follows: Sect. 3.2 reviews loop-tiling fundamentals; Sect. 3.4 details the mCAKE analytical model; Sect. 3.5 specializes the model to direct convolutions; Sects. 3.6.3 and 3.6.4 present experimental results.

3.2 Background and Notation

This section reviews the two core kernel families targeted by mCAKE, namely tensor contractions and direct convolutions. It also establishes the notation used throughout the chapter. A brief recap of matrix multiplication is included only insofar as it clarifies terminology.

3.2.1 Matrix Multiplication (Recap)

A matrix multiplication $C \leftarrow C + AB$ with $A \in \mathbb{R}^{M\times K}$ and $B \in \mathbb{R}^{K\times N}$ performs MKN multiply–accumulate (MAC) operations. Conceptually this is a three–loop nest over indices (i, j, k). *Tiling* partitions the iteration space into $m \times n \times k$ *computation blocks*, each operating on tiles of A, B, and C that fit in on-chip memory. Data held for reuse is called *stationary*; data streamed from or to DRAM exactly once per block is called *streaming*.

3.2.2 Tensor Contractions

Tensor contraction generalizes GEMM to higher ranks by reducing over one or more paired indices. For example,

$$C[a, b, c, d] = \sum_{k} A[a, b, k]\ B[k, c, d]$$

is a six-loop computation (three in A, three in B) that reduces over k. In Einstein notation we write `abcd-kcd-abk`. The search space for loop orders and tile sizes grows combinatorially with tensor rank, motivating an analytic schedule such as mCAKE.

3.2.3 Direct Convolution vs. GEMM-Convolution

Convolution is an operation prevalent in machine learning for extracting features of images. A convolution layer works by sliding and applying filters over the input data to detect patterns. For example, a single convolution filter might detect vertical lines when applied to an input image.

For a 2D image, each filter is a 3D tensor that is multiplied element-wise with many input patches. In order to take advantage of existing high-performance GEMM kernels [27], the convolution operation is typically computed as an matrix multiplication between a weight and data matrix [22]. Traditional im2col incurs data replication linearizing overlapping patches. Elements that are a member of multiple patches will be duplicated in the linearized data matrix. Figure 3.2 shows an example of this data replication.

The replication results in an expanded data matrix in the off-chip DRAM memory which increases DRAM bandwidth usage since the same values are copied multiple times. In addition, duplication may also decrease the effective usable cache size. Instead of reusing an element in cache multiple times, elements are replicated once for every patch that they are a member of. Particularly, on resource constrained devices, such as those used in TinyML, wasting memory bandwidth and cache space with duplicated values will usually reduce computation throughput and limit model sizes.

To avoid data duplication, techniques such as indirect convolution [9, 31] have been proposed. More generally, methods such as direct convolution [30] implement convolution as a set of nested loops (shown in Algorithm 3). Using mCAKE we can generally optimize schedules for sets of nested loops, such as the 6 nested loops of direct convolution to

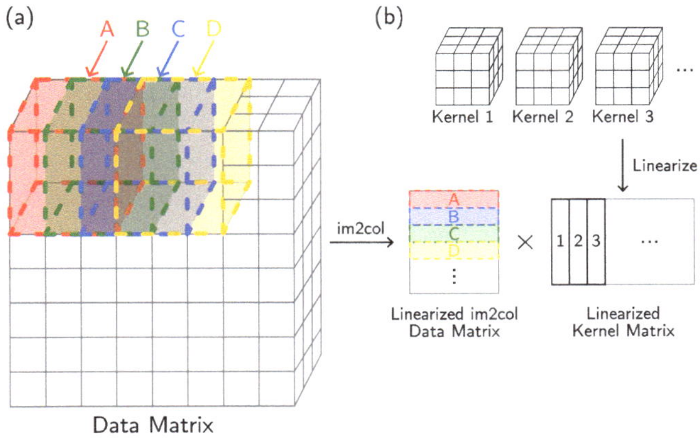

Fig. 3.2 (**a**) 3D input data is linearized into a data matrix via im2col. im2col creates a vector for each patch (labeled A-D). When the stride size is less than the kernel size patches will overlap. Values belonging to multiple patches are replicated in the data matrix. In this example, elements in the middle of the input data tensor belong to three patches and are therefore replicated 3 times in the data matrix. (**b**) the weight matrix is formed by linearizing each kernel (filter) into a vector which is compatible with the im2col linearization in (**a**)

decrease off-chip memory accesses. That is, mCAKE selects the 3 innermost loops to maximize on-chip data reuse and then analytically derives tile sizes for the 3 innermost loops.

3.3 Related Work

3.3.1 Vendor Libraries and Hand–Tuned Kernels

Commercial libraries such as Intel MKL and oneDNN offer highly optimized GEMM, convolution, and tensor–contraction routines [1,13]. Their speed derives from architecture-specific micro-kernels and an extensive catalogue of schedules and data layouts. Scaling to many cores, however, typically follows the Goto–style strategy of replicating tiles among threads, thereby *increasing off-chip DRAM traffic in proportion to the core count* [10]. On bandwidth-constrained systems this yields sub-linear speed-up and significant energy cost [16]. mCAKE instead fixes a constant DRAM traffic budget and exploits on-chip cache bandwidth which is sufficient on modern CPUs. Pure C code is generated, avoiding hand-written assembly.

3.3.2 Tile-Size Search and Analytical Cost Models

Recent frameworks narrow the search space of loop permutations and tile sizes via analytical models, then call a linear or MILP solver to pick tiles at each cache level (e.g. ACMTC [19] and HPCRL [20]). Because the solver is invoked hierarchically, optimality at one level does not guarantee global optimality, and in practice some of the generated schedules *increase* DRAM traffic (Sect. 3.6.3). Moreover, cache bandwidth is rarely the bottleneck on current CPUs; a model that minimizes cache traffic can under-utilize the very resource that is most abundant. mCAKE sidesteps this issue by targeting the single scarcest resource—off-chip bandwidth—and derives tile sizes analytically.

3.3.3 Polyhedral Compilation

Polyhedral compilers (e.g. Polly [11], Pluto [7], MLIR affine [18], Tiramisu [5]) express loop nests as integer polyhedra and apply affine transforms to improve locality and parallelism. Although matrix multiplication, tensor contractions, and direct convolutions are affine, choosing tile sizes that meet a quantitative bandwidth budget is *non-linear* and thus falls outside the pure affine framework. Consequently, state-of-the-art polyhedral compilers resort to auto-tuning or grid search, often with considerable programmer intervention, yet still lag vendor libraries in speed. By contrast, mCAKE achieves competitive or better performance *without* any design-space exploration and with minimal programmer effort.

3.4 mCAKE: Extending CAKE to Deep Loop Nests

Figure 3.3 gives a visual summary of the multi-CAKE (mCAKE) pipeline. Starting from an n-deep loop nest (Fig. 3.3a) we:

- **Pick the three innermost loops** to maximize data reuse (Sect. 3.4.1).
- **Apply CAKE** to those loops, obtaining a constant-bandwidth *block shape* that fits in the on-chip memory (Sect. 3.4.2).
- **Permute the three loops** to minimize total DRAM traffic for that block (Sect. 3.4.3).
- Treat the remaining outer loops as *instance generators*: each outer iteration launches a new CAKE-optimized inner kernel (Fig. 3.3f, g).

The result is a fully analytic schedule that delivers constant or near-constant DRAM bandwidth, yet requires *no* auto-tuning.

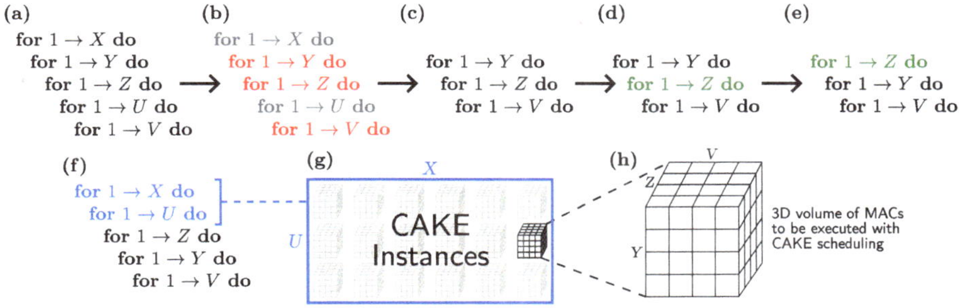

Fig. 3.3 mCAKE workflow. (**a**) Original n-deep loop nest with reduction V. (**b**) Select three reuse-critical loops. (**c–e**) Permute to make them innermost and apply CAKE-tiling. (**f** and **g**) Outer loops replicate the CAKE kernel. (**h**) Resulting constant-bandwidth computation blocks

3.4.1 Selecting the Three Innermost Loops

The reduction index is the only dimension that both *reads* inputs *and writes* partial results; keeping its tile stationary therefore offers the largest reuse opportunity. We thus require that *at least one* of the innermost three loops be a reduction dimension. Among the remaining indices we choose the largest one as the second innermost loop so that the stationary surface spans two large dimensions, maximizing reuse. The third innermost loop is the largest of the leftovers; its tiles are streamed once per block.

3.4.2 Tile Size via CAKE

With the loop triple fixed, we invoke CAKE's *constant-bandwidth blocking* to derive an on-chip block of size $m \times n \times k$ (Fig. 3.3h). To break the cyclic dependency between loop order and tile size we *temporarily* assume that the reduction surface (partial result tile) is the stationary one; this yields a valid CAKE block shape for any later permutation of the three loops. If the dimensions are severely skewed the block may not be absolutely optimal, but Sect. 3.4.3 compensates by picking the loop order that minimizes global traffic.

3.4.3 Choosing the Optimal Inner-Loop Order

For a candidate permutation (i, j, k) of the three innermost loops, total DRAM traffic is

$$\mathrm{IO}(i, j, k) = \underbrace{(\#\mathrm{blocks})}_{\prod_{\ell \notin \{i,j,k\}} P_\ell} \times \underbrace{(\mathrm{tile}_j + \mathrm{tile}_k)}_{\text{stream}} + \underbrace{\mathrm{surface}_{i,j}}_{\text{stationary}}, \tag{3.1}$$

where P_ℓ is the loop bound of outer index ℓ, tile_j is the size of the streamed tile along loop j, and $\text{surface}_{i,j}$ is the size of the stationary $i \times j$ surface.[1] We evaluate (3.1) for all six permutations and select the order with minimal traffic. Section 2.5.1 and Fig. 2.6 demonstrate how this choice yields the highest throughput for both $M \gg N, K$ (choose M-first) and $K \gg M, N$ (choose K-first) scenarios.

The outer loops (4th and beyond) merely spawn CAKE instances; their order affects only a small fraction of the runtime and is left unchanged.

3.4.4 Outer–Loop Ordering and Memory-Access Linearity

Once the CAKE-optimized inner loop is fixed, opportunities for temporal reuse across *outer* iterations are minimal. Consequently, the choice of outer-loop permutation is driven less by reuse and more by *how* the tensor is laid out in memory. Ordering outer loops so that the fastest-varying index in memory is traversed innermost maximizes spatial locality and reduces TLB pressure, whereas reversing that order can provoke strided or scattered accesses that waste pre-fetch bandwidth and inflate latency. In practice we therefore align the outer-loop order with the physical storage order of the operand that dominates streaming traffic (e.g. row-major vs. column-major), a consideration that outweighs the marginal reuse available beyond the three innermost loops.

3.4.5 Summary of mCAKE

mCAKE inherits CAKE's constant-bandwidth guarantee while generalizing it to arbitrary tensor contractions and direct convolutions. The method requires no search: loop-triple selection is rule-based, the CAKE tile shape is analytic, and among six permutations the IO-optimal order is chosen via Eq. (3.1). Subsequent sections validate this schedule experimentally on Intel and ARM CPUs.

3.5 mCAKE for Direct Convolution

Direct (six-loop) convolution differs from GEMM in that an input element may participate in many overlapping patches when the stride is less than the kernel size. Exploiting this reuse is critical to improving on im2col-based GEMM, which duplicates such elements in DRAM (Fig. 3.2). We therefore treat direct convolution as a 6-deep loop nest and apply mCAKE exactly as in Sect. 3.4.

[1] A read and a write are counted as two accesses.

3.5.1 Loop Nest and CAKE Mapping

Algorithm 3 shows the naïve nesting (m, k, h, w, i, j). mCAKE chooses the three innermost loops as (k, i, j) (large reduction space) and tiles them into constant–bandwidth $m_t \times k_t \times n_t$ blocks (Algorithm 4). The outer loops (h, m) spawn CAKE kernels across the output spatial grid and filter bank.

Algorithm 3: Original loops for direct convolution

Input:

- $C_{out} \times C_{in} \times H_f \times W_f$ filter F
- $C_{in} \times H_{in} \times W_{in}$ input feature map In
- stride s

Output: $C_{out} \times H_{out} \times W_{out}$ output feature map Out

for $m = 0$ **to** C_{out} **do**
  for $k = 0$ **to** C_{in} **do**
    for $h = 0$ **to** H_{out} **do**
      for $w = 0$ **to** W_{out} **do**
        for $i = 0$ **to** H_f **do**
          for $j = 0$ **to** W_f **do**
            $Out[m][h][w] = In[k][h * s + i][w * s + j] \times F[m][k][i][j]$

Algorithm 4: mCAKE K-first direct convolution after tiling and loop reordering

---

Input:

- $C_{out} \times C_{in} \times H_f \times W_f$ filter F
- $C_{in} \times H_{in} \times W_{in}$ input feature map In
- number of $m_t \times k_t \times n_t$ CB blocks in each dimension
  $M_b = C_{out}/m_t$; $K_b = C_{in}/k_t$; $H_b = H_{out}/n_t$
- stride s

Output: $C_{out} \times H_{out} \times W_{out}$ output feature map Out

for $h = 0$ **to** H_b **do**
  for $m = 0$ **to** M_b **do**
    for $k = 0$ **to** K_b **do**
      `// parallelized CB block multiplication`
      for $n = 0$ **to** n_t **do**
        for $i = 0$ **to** H_f **do**
          for $j = 0$ **to** W_f **do**
            for $nn = 0$ **to** W_{out}/n_r **do**
              for $mm = 0$ **to** m_t/m_r **do**
                `//` $m_r \times n_r$ `outer product`
                (outer-product between the input and filter tiles)

3.5.2 DRAM Traffic for Direct Convolution

For a constant-bandwidth (CB) block of shape $(pC_m) \times (H_f W_f C_k) \times (pC_m)$ (first two dimensions stream, third is stationary), the off-chip accesses are

$$\mathrm{IO}_{\mathrm{direct}} = pC_m\, C_k\big(H_f W_f + 1\big).$$

Dividing by compute time $T = \mathrm{MACs}/(pf) = (pC_m)^2 H_f W_f C_k/(pf)$ gives a bandwidth requirement that *does not grow with the core count* p:

$$BW_{\mathrm{direct}} \approx \frac{f}{C_m}.$$

3.5.3 Why GEMM Convolution Costs More Bandwidth

In im2col GEMM each input value is replicated up to $H_f W_f$ times. Let C'_m, C'_k be the reduced tile sizes that still fit in L2 after replication. The CB block now performs $(pC'_m)^2 H_f W_f C'_k$ MACs but must read *two* replicated tiles:

$$\mathrm{IO}_{\mathrm{gemm}} = 2\, pC'_m\, C'_k H_f W_f \quad \Longrightarrow \quad BW_{\mathrm{gemm}} \approx \frac{2f}{C'_m}.$$

Thus GEMM convolution needs $\geq 2\times$ the off-chip bandwidth of direct convolution and provides less FLOPs per byte because replication shrinks the usable tile.

Take-Away mCAKE retains CAKE's constant-bandwidth guarantee for stride-1 direct convolution: scaling to more cores increases compute throughput without raising DRAM traffic, whereas GEMM-convolution remains bandwidth-bound and duplicative.

3.6 Performance Evaluation on CPUs

We benchmark mCAKE on tensor contraction and direct convolution workloads across two architecturally distinct CPUs, measuring on-chip throughput (GFLOP/s) and off-chip DRAM bandwidth (GB/s). Our goal is to show that mCAKE sustains near-peak computation while holding DRAM traffic constant.

Table 3.1 CPUs used in mCAKE evaluation

CPU	L1	L2	L3	DRAM	Cores	LLC BW	DRAM BW
Intel i9-10900K	32 KiB	256 KiB	20 MiB	32 GB	10	225 GB/s	40 GB/s
ARM Cortex-A72	32 KiB	1MiB	N/A	8 GB	4	35 GB/s	3.5 GB/s

3.6.1 CPU Platforms

Table 3.1 summarizes the two test beds. Simultaneous multithreading (SMT) and dynamic frequency and voltage scaling (DVFS) are disabled, and caches are flushed between runs. Results are averaged over 50 trials.

- **ARM Cortex-A72**. A low-power quad-core with very limited DRAM bandwidth but ample cache bandwidth—representative of embedded systems.
- **Intel Core i9-10900K**. Ten cores, large caches, and higher DRAM bandwidth, yet still compute-heavy relative to memory.

3.6.2 Implementation Details

Tensor-contraction and convolution kernels are written in C++ with ISA-specific intrinsics. The inner three loops invoke the original CAKE kernel; no hand-tuned assembly is used. All packing or data-layout transforms (including `im2col`) are timed and counted in bandwidth numbers.

3.6.3 Tensor Contractions

We benchmark mCAKE against ACMTC [19] and TCL [25] on the various tensor contractions in TCCG suite [26] using a single core of the Intel i9-10900K.

Dimensions for each contraction appear in Table 3.2.

Baselines ACMTC employs the Couenne MINLP solver [6] to pick architecture-specific tile sizes at each cache level; we supply the exact cache capacities and bandwidths of our test CPU. TCL performs explicit tensor transposes followed by MKL GEMM calls [1]. All layout transforms are included in the timing and DRAM-traffic measurements collected with Intel VTune [14].

Observation ACMTC's hierarchical optimisation targets latency at each cache level independently; DRAM traffic is not minimised globally. mCAKE, in contrast, shapes the *three* reuse-critical loops to guarantee constant (and minimal) off-chip traffic, leading to

Table 3.2 Tensor contraction problem sizes from the TCCG [26] benchmark used in Fig. 3.4. We represent the problems using Einstein notation: output-input$_2$-input$_1$ and give the size of each dimension on the right

Expression	Problem size
abc-acd-db	a:312 b:312 c:312 d:312
abcd-ebad-ce	a:72 b:72 c:24 d:72 e:72
ab-acd-dbc	a:312 b:312 c:312 d:312
abcde-ecbfa-fd	a:48 b:32 c:32 d:24 e:48 f:48
abcd-aedf-bfce	a:72 b:72 c:72 d:72 e:72 f:72

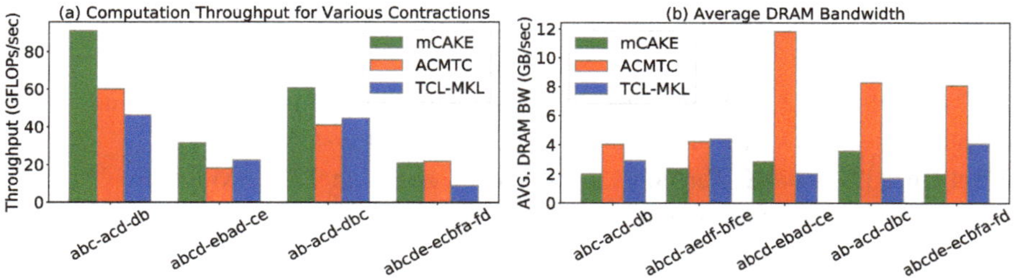

Fig. 3.4 Performance comparisons on Intel i9-10900K system for tensor contraction computations. (**a**) Single-core computation throughput and (**b**) DRAM bandwidth of mCAKE, TCL [25], and ACMTC [19] on the TCCG [25] tensor contraction benchmark measured on an Intel Core i9-10900K system. Table 3.2 shows the dimension sizes for each of the evaluated tensor contraction problems. When the input tensors are larger, mCAKE performs better because inputs may not fit in the CPU cache and minimizing off-chip DRAM memory accesses becomes critical

higher GFLOP/s and markedly lower bandwidth. As Fig. 3.4b shows, ACMTC consumes up to 2×2× the DRAM bandwidth of mCAKE, while mCAKE sustains the highest throughput across all TCCG contractions.

3.6.4 Direct Convolution Performance

We tested a CNN benchmark from HPCRL [21] containing 16 2D convolution operations from ResNet-18 [12] and Yolo-9000 [24] networks. We show the specific layer configurations in Table 3.3. On the Intel CPU, we compared mCAKE to Intel oneDNN [13], Intel MKL [1], and HPCRL [20] and use the VTune Profiler [14] to measure throughput and average DRAM bandwidth. HPCRL extends ACMTC's method of searching for tile sizes to direct convolution in a multi-core setting. On the ARM CPU, we compare mCAKE to ARM Compute Library's GEMM-based convolution [4] and collect performance data using the Linux perf Tool [2]. We use Perf to record DRAM accesses by monitoring the ARM PMU event counter for L2 cache refills from DRAM.

Table 3.3 Configuration of conv2d layers

Layer	C_{out}	C_{in}	H_{in}/W_{in}	R/S
Yolo2	64	32	272	3
Yolo4	128	64	136	3
Yolo5	64	128	136	1
Yolo8	256	128	68	3
Yolo9	128	256	68	1
Yolo12	512	256	34	3
Yolo13	256	512	34	1
Yolo18	1024	512	17	3
Yolo23	1024	1024	17	1

Layer	C_{out}	C_{in}	H_{in}/W_{in}	R/S
Resnet2	64	64	56	3
Resnet3	64	64	56	1
Resnet6	128	128	28	3
Resnet8	256	128	28	3
Resnet9	256	256	14	3
Resnet12	512	512	7	3

With GEMM-based convolution, im2col replication results in an expanded data matrix in the off-chip memory, increasing DRAM bandwidth usage (ARMCL in Fig. 3.5c). mCAKE uses direct convolution so we load data shared by multiple patches only once, i.e., reuse the loaded data across multiple overlapping patches. This allows mCAKE to achieve high performance relative to GEMM-based convolution routines in Fig. 3.5a (MKL GEMM). Since the Cortex-A72 CPU has constrained DRAM bandwidth resources (see Table 3.1), mCAKE outperforms ARMCL on every benchmark (Fig. 3.5b) while also using considerably less DRAM bandwidth (Fig. 3.5c).

On the Intel CPU, mCAKE outperforms or matches the performance of oneDNN's direct convolution for 14 out of the 16 CNN layers in the benchmark (Fig. 3.5a). However for certain layers where inputs fit entirely in the CPU's cache (e.g., Yolo9, Resnet2 in Table 3.3), mCAKE performs worse than oneDNN. Since the Intel CPU has large cache capacity and DRAM bandwidth resources, mCAKE attains smaller speedup gains from DRAM bandwidth minimization relative to the ARM CPU (Fig. 3.6).

3.6.5 Throughput and Bandwidth Usage When Scaling Core Count

In this section, we demonstrate the performance scaling of mCAKE by fixing the problem size and growing the number of cores while measuring throughput and DRAM bandwidth usage (Fig. 3.7). We use the Intel i9-10900K CPU and perform a fixed direct convolution with 3×3 filters, a 32×32 input image, 512 input and output channels, and a stride of 1. Both CAKE and oneDNN performance in computation throughput scale with the number of cores (Fig. 3.7a) but mCAKE does not need to increase DRAM bandwidth to utilize more cores (Fig. 3.7b).

In Fig. 3.7c and d, we profile the 'abc-acd-db' tensor contraction problem with dimensions $a = b = 312$ and $c = d = 1024$. ACMTC does not tile the computation for parallelism in the innermost three dimensions. Instead, ACMTC scales performance in multi-core settings by running a several independent tensor contraction problems in parallel (batch). By contrast, mCAKE can exploit parallelism in the 3 innermost

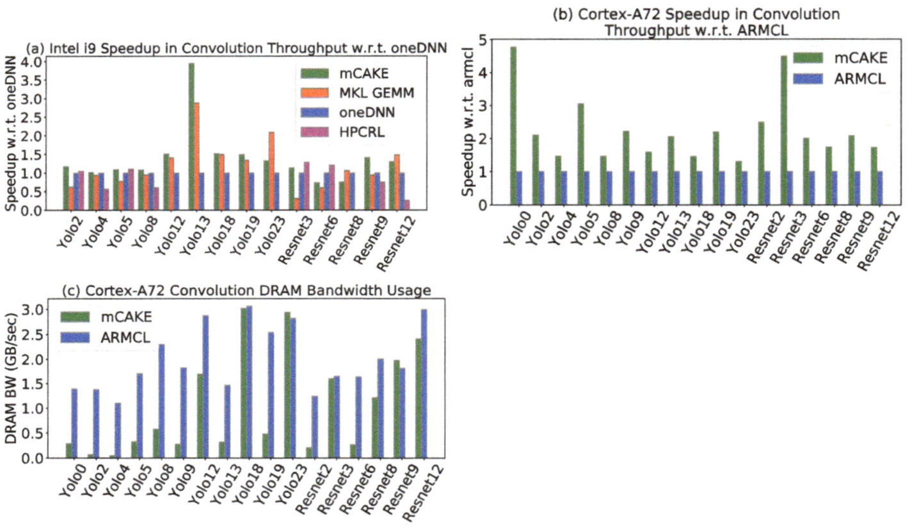

Fig. 3.5 Performance comparisons for convolution computation. (**a**) Speedup in computation throughput relative to Intel MKL [1], oneDNN [13], and HPCRL [20] on an Intel Core i9-10900K system, (**b**) speedup in computation throughput relative to ARMCL [4] on an Arm Cortex-A72 based system, and (**c**) DRAM bandwidth usage for the ARM system. Results are for all unique convolution layers in Yolo-9000 [24] and ResNet-18 [12]. Table 3.3 shows the configurations for each layer. We include all data layout transformations (e.g., im2col) when measuring performance

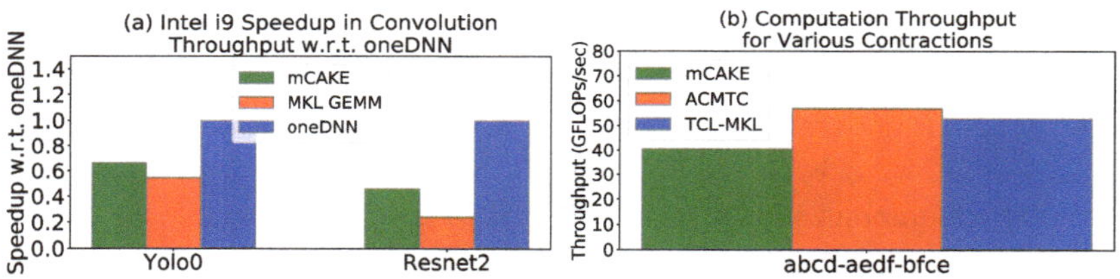

Fig. 3.6 mCAKE's performance on direct convolution and tensor contraction problems where some or all inputs fit entirely in CPU caches

dimensions of each contraction, as well as across a batch, and scale performance with the number of cores (Fig. 3.7c) while holding DRAM bandwidth nearly constant (Fig. 3.7d). We omit performance results for the ARM Cortex-A72 system as it only contains 4 cores.

3.7 Summary

This chapter presented **mCAKE**, a fully-analytic scheduling framework that extends CAKE's constant-bandwidth principles from three-loop GEMM to deep loop nests that

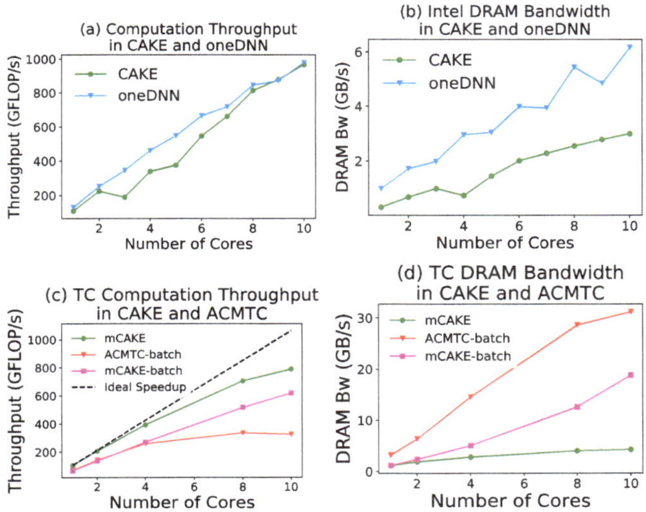

Fig. 3.7 Computation throughput and DRAM bandwidth usage of mCAKE and competitors for a fixed direct convolution (**a**,**b**) and tensor contraction (**c**,**d**) problem when increasing the number of cores on the Intel i9 CPU. Both CAKE and oneDNN performance scale with the number of cores (**a**) but CAKE does not increase DRAM bandwidth usage to utilize more cores (**b**). ACMTC can perform independent tensor contractions in parallel (batch), but cannot tile the computation for parallelism in the 3 innermost dimensions. By contrast, CAKE can exploit parallelism in the 3 innermost loops as well as a batch of tensor contractions and scale performance with the number of cores (**c**) while holding DRAM bandwidth nearly constant (**d**)

include a reduction dimension (e.g. tensor contractions and direct convolutions). The method is deliberately simple:

1. *Choose* as innermost the reduction loop and the two largest remaining loops, maximizing on-chip reuse;
2. *Apply* the CAKE constant-bandwidth tiling to this three-loop kernel;
3. *Spawn* CAKE kernels across the outer loops without any further search or tuning.

Despite its simplicity, mCAKE matches or exceeds the performance of state-of-the-art search-based systems. On both an embedded Cortex-A72 and a desktop Core i9, it sustains near-peak GFLOP/s while holding DRAM traffic constant and often significantly less than the baselines. These results demonstrate that an analytically derived, bandwidth-aware schedule can outperform heavyweight design-space exploration, offering a practical path toward memory-wall mitigation on modern multi-core CPUs.

References

1. *Intel oneAPI Math Kernel Library.*
2. perf: Linux profiling with performance counters. https://perf.wiki.kernel.org/index.php/Main_Page, 2021.
3. John R. Allen and Ken Kennedy. Automatic loop interchange. In *Proceedings of the 1984 SIGPLAN Symposium on Compiler Construction*, SIGPLAN '84, page 233–246, New York, NY, USA, 1984. Association for Computing Machinery.
4. Arm Limited. *Arm Compute Library Reference Guide*, January 2022.
5. Riyadh Baghdadi, Jessica Ray, Malek Ben Romdhane, Emanuele Del Sozzo, Abdurrahman Akkas, Yunming Zhang, Patricia Suriana, Shoaib Kamil, and Saman Amarasinghe. Tiramisu: A polyhedral compiler for expressing fast and portable code. In *Proceedings of the 2019 IEEE/ACM International Symposium on Code Generation and Optimization*, CGO 2019, pages 193–205, Washington, DC, USA, February 2019. IEEE Press.
6. Pietro Belotti. Couenne: a user's manual, 2016.
7. Uday Bondhugula, Albert Hartono, J. Ramanujam, and P. Sadayappan. A practical automatic polyhedral parallelizer and locality optimizer. In *Proceedings of the 29th ACM SIGPLAN Conference on Programming Language Design and Implementation*, PLDI '08, page 101–113, New York, NY, USA, 2008. Association for Computing Machinery.
8. Tianqi Chen, Thierry Moreau, Ziheng Jiang, Lianmin Zheng, Eddie Yan, Haichen Shen, Meghan Cowan, Leyuan Wang, Yuwei Hu, Luis Ceze, Carlos Guestrin, and Arvind Krishnamurthy. TVM: An automated End-to-End optimizing compiler for deep learning. In *13th USENIX Symposium on Operating Systems Design and Implementation (OSDI 18)*, pages 578–594, Carlsbad, CA, October 2018. USENIX Association.
9. Marat Dukhan. The indirect convolution algorithm, 2019.
10. Kazushige Goto and Robert A. van de Geijn. Anatomy of high-performance matrix multiplication. *ACM Trans. Math. Softw.*, 2008.
11. Tobias Grosser, Groesslinger Armin, and Christian Lengauer. Polly - performing polyhedral optimizations on a low-level intermediate representation. *Parallel Processing Letters*, 22(04), December 2012.
12. Kaiming He, Xiangyu Zhang, Shaoqing Ren, and Jian Sun. Deep residual learning for image recognition, 2015.
13. Intel Corporation. *Intel oneAPI Deep Neural Network Library*, March 2016.
14. Intel Corporation. *Intel VTune Profiler*, 2021.
15. Andrei Ivanov, Nikoli Dryden, Tal Ben-Nun, Shigang Li, and Torsten Hoefler. Data movement is all you need: A case study on optimizing transformers, 2020.
16. H. T. Kung, Vikas Natesh, and Andrew Sabot. Cake: Matrix multiplication using constant-bandwidth blocks. In *Proceedings of the International Conference for High Performance Computing, Networking, Storage and Analysis*, SC '21. Association for Computing Machinery, 2021.
17. Monica D. Lam, Edward E. Rothberg, and Michael E. Wolf. The cache performance and optimizations of blocked algorithms. page 63–74, 1991.
18. Chris Lattner and Jacques Pienaar. MLIR primer: A compiler infrastructure for the end of moore's law, 2019.
19. Rui Li, Aravind Sukumaran-Rajam, Richard Veras, Tze Meng Low, Fabrice Rastello, Atanas Rountev, and P. Sadayappan. Analytical cache modeling and tilesize optimization for tensor contractions. In *Proceedings of the International Conference for High Performance Computing, Networking, Storage and Analysis*, SC '19, New York, NY, USA, 2019. Association for Computing Machinery.

20. Rui Li, Yufan Xu, Aravind Sukumaran-Rajam, Atanas Rountev, and P. Sadayappan. Analytical characterization and design space exploration for optimization of CNNs. In *Proceedings of the 26th ACM International Conference on Architectural Support for Programming Languages and Operating Systems*, ASPLOS '21, page 928–942, New York, NY, USA, 2021. Association for Computing Machinery.
21. Rui Li, Yufan Xu, Aravind Sukumaran-Rajam, Atanas Rountev, and P. Sadayappan. Analytical characterization and design space exploration for optimization of cnns. In *Proceedings of the 26th ACM International Conference on Architectural Support for Programming Languages and Operating Systems*, ASPLOS '21, page 928–942, New York, NY, USA, 2021. Association for Computing Machinery.
22. Bradley McDanel, Sai Qian Zhang, HT Kung, and Xin Dong. Full-stack optimization for accelerating CNNs using powers-of-two weights with FPGA validation. In *Proceedings of the ACM International Conference on Supercomputing*, pages 449–460, 2019.
23. Kathryn S. McKinley, Steve Carr, and Chau-Wen Tseng. Improving data locality with loop transformations. *ACM Trans. Program. Lang. Syst.*, 18(4):424–453, jul 1996.
24. Joseph Redmon, Santosh Divvala, Ross Girshick, and Ali Farhadi. You only look once: Unified, real-time object detection, 2015.
25. Paul Springer and Paolo Bientinesi. Design of a high-performance GEMM-like tensor-tensor multiplication, 2016.
26. Paul Springer and Paolo Bientinesi. Design of a high-performance GEMM-like Tensor-Tensor Multiplication. *CoRR*, 2016.
27. Pete Warden. Why GEMM is at the heart of deep learning, April 2015.
28. Wikipedia contributors. ARM cortex-a — Wikipedia, the free encyclopedia. https://en.wikipedia.org/w/index.php?title=ARM_Cortex-A&oldid=1096388644, 2022. [Online; accessed 19-October-2022].
29. Wikipedia contributors. Xeon — Wikipedia, the free encyclopedia. https://en.wikipedia.org/w/index.php?title=Xeon&oldid=1108807552, 2022. [Online; accessed 19-October-2022].
30. Jiyuan Zhang, Franz Franchetti, and Tze Meng Low. High performance zero-memory overhead direct convolutions. In Jennifer Dy and Andreas Krause, editors, *Proceedings of the 35th International Conference on Machine Learning*, volume 80 of *Proceedings of Machine Learning Research*, pages 5776–5785. PMLR, 10–15 Jul 2018.
31. Yangjie Zhou, Mengtian Yang, Cong Guo, Jingwen Leng, Yun Liang, Quan Chen, Minyi Guo, and Yuhao Zhu. Characterizing and demystifying the implicit convolution algorithm on commercial matrix-multiplication accelerators, 2021.

Rosko: Sparse Matrix Multiplication for Machine Learning Workloads

4

Chapter Overview

The previous chapters established a foundation for optimizing dense matrix and tensor computations through hardware-aware tiling and scheduling. However, a significant portion of the "computational tax" in modern AI stems from redundant operations on zero-valued parameters. In this chapter, we turn our attention to the challenge of sparsity as it arises in deep learning workloads.

This chapter introduces Rosko, a method designed to efficiently handle the particular patterns of sparsity found in neural networks. The name "Rosko" stands for Row Skipping Outer products, capturing the core idea of efficiently skipping computation when zeros are encountered during outer product matrix multiplication. Building on the ideas of CAKE discussed in Chap. 2 and its generalization to multi-dimensional tensors, we extend the core techniques to leverage sparsity in a hardware-friendly way. Rather than relying on costly indexing, Rosko exploits the structure of outer products by enabling the skipping of entire rows of computation when zeros are encountered, yet doing so with minimal overhead. Since Rosko only skips zero multiplications there is no change to the result of the computation. The method is broadly applicable across different hardware platforms.

We also demonstrate an important synergy between network pruning and hardware efficiency: by aligning the sparsity pattern of pruned neural networks with the hardware's memory access granularity (e.g., aligning blocks of 16 dense elements), we can ensure that the computational workload is efficiently distributed, even at moderate sparsity levels. For example, with 50% sparsity, Rosko enables pruning in such a way that each group of 32 elements contains exactly 16 nonzeros, maximizing throughput on hardware architectures that move data in blocks of 16.

The following chapter will explore these ideas: starting with practical motivation and background, and proceeding through implementation, hardware considerations, and

A. Sabot, *Minimizing Data Movement and Parameter Count Across the Machine Learning Stack*, Synthesis Lectures on Computer Science,
https://doi.org/10.1007/978-3-032-23100-0_4

empirical results showcasing how Rosko is able to take advantage of moderate sparsity in modern deep learning applications.

4.1 Introduction to Rosko

The surge in deep learning adoption, particularly through architectures such as convolutional neural networks (CNNs) and transformers, has led to matrix multiplication (MM) becoming a core operation for modern artificial intelligence workloads. As these models grow in size, reducing their computational and memory requirements is critical, not only to extend their reach to everyday devices but also to accelerate training and inference on current hardware.

A promising approach to increasing efficiency is leveraging sparsity: by eliminating redundant or insignificant weights, we can reduce the number of computations and memory accesses. However, exploiting sparsity in practice is challenging. Unstructured pruning, which removes individual weights indiscriminately, yields high accuracy but often leads to irregular data access patterns and high overheads when executed on traditional hardware, negating many of the performance benefits. On the other hand, structured pruning produces more regular patterns but typically at the expense of model accuracy.

In this chapter, we introduce Rosko, a collection of sparse matrix multiplication (SpMM) kernels that enable both computational and memory efficiency for deep neural networks (DNN). Rosko leverages row skipping in outer product computations: during execution, entire rows corresponding to zeroed input values can be bypassed, greatly reducing the number of unnecessary calculations. What distinguishes Rosko from conventional sparse computation frameworks is its ability to achieve both fine-grained (for accuracy) and structured (for hardware efficiency) sparsity through a combination of row skipping-aware pruning and training.

Crucially, Rosko avoids the indexing and management overhead typical of sparse formats like CSR or CSC, thanks to a packed data layout that streamlines memory accesses even at moderate sparsity levels. Rosko's design is compatible with established scheduling approaches such as CAKE (Chap. 2) and Goto's algorithm [9], both of which maximize data reuse in dense matrix multiplication. This compatibility helps Rosko bridge the gap between dense and sparse libraries.

A key motivation for Rosko is the practical performance gap between existing dense and sparse matrix multiplication libraries, particularly in the intermediate sparsity regime relevant for pruned neural networks deployed on real hardware. To demonstrate this, Fig. 4.1 compares the wall-clock runtimes of Rosko kernels to both dense and sparse libraries across a range of sparsity levels on two representative CPUs.

As seen in Fig. 4.1, Rosko achieves substantial runtime reductions over both dense and sparse matrix multiplication kernels within a wide range of sparsities on both Intel and ARM CPUs. These results highlight the effectiveness of Rosko's approach, particularly

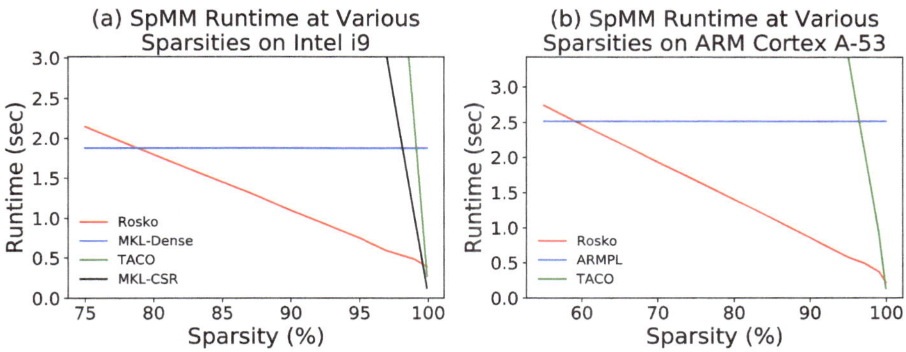

Fig. 4.1 Wall-clock SpMM runtime for Rosko kernels compared to sparse and dense libraries. Rosko outperforms dense MM and SpMM kernels from (**a**) 75% to 99.5% sparsity on Intel and (**b**) 58% to 99.5% sparsity on ARM CPUs

its ability to target the intermediate sparsity regime that is underserved by conventional libraries, and constitute a primary motivation for the work presented in this chapter.

On real-world CPUs, Rosko outperforms leading dense and sparse libraries, including Intel's oneMKL and ARM Compute Library, across a wide range of sparsities. For example, at 95% sparsity, Rosko kernels achieve up to a 6.5× reduction in runtime on Intel CPUs and a 4.7× reduction on ARM CPUs.

In summary, this chapter makes the following contributions:

- Introduces the Rosko algorithm for efficient sparse MM using row skipping and packed kernel design, reducing both computation and sparsity management overhead.
- Proposes row skipping-aware training and iterative pruning, producing high-accuracy models with hardware-friendly structured sparsity.
- Provides efficient Rosko library implementations that outperform existing CPU SpMM and dense MM libraries across practical scenarios, demonstrating Rosko's effectiveness for a wide range of sparsities.

Through these innovations, Rosko enables new levels of matrix multiplication efficiency, making it easier to harness the power of sparsity in modern neural networks.

4.2 Background

In this section, we provide the background necessary for understanding Rosko's contributions. We first situate Rosko kernels in the landscape of existing dense and sparse matrix multiplication libraries, and then introduce the outer product approach for efficient matrix multiplication computation.

4.2.1 Dense and Sparse Matrix Multiplication Kernels

Matrix multiplication arises in various forms, such as between two dense matrices, between a dense and a sparse matrix, or between two sparse matrices. Well-known libraries—including OpenBLAS [22], BLIS [21], and Intel oneMKL [13]—provide highly optimized routines for dense-dense multiplication. Sparse matrix multiplication (SpMM), where at least one input is sparse, is addressed by specialized kernels within these and other libraries. For sparse-sparse matrix multiplication, both operands are sparse.

However, as demonstrated in Fig. 4.1, existing sparse matrix multiplication libraries typically outperform dense libraries only at very high sparsities (i.e., when most elements are zero). For example, Intel oneMKL's SpMM routine only surpasses dense performance when sparsity exceeds 99%. For lower levels of sparsity—a regime of practical interest in many pruned neural networks—sparsity management overheads dominate, and dense routines remain more efficient. This intermediate sparsity regime, identified in [8], is often neglected by both dense and sparse libraries, leaving performance gains unrealized, especially on CPUs and GPUs.

4.2.2 Outer Products for Matrix Multiplication to Increase Arithmetic Intensity

Outer product-based matrix multiplication offers a route to higher arithmetic intensity—the ratio of computations performed to external memory bandwidth consumed—than conventional inner product-based matrix multiplication, through more efficient data streaming. Improving arithmetic intensity is critical for better utilization of limited memory bandwidth.

Consider the standard matrix multiplication operation $C = A \times B$. This computation can be decomposed into smaller subproblems, each responsible for computing a tile of the output matrix C. Specifically, by tiling A and B into submatrices of size $m \times k$ and $k \times n$, we can break C into $m \times n$ tiles. Each tile can then be computed via inner or outer product formulations, as illustrated in Fig. 2.2.

To maximize the arithmetic intensity of the overall matrix multiplication, it is necessary to do the same for each tile. Inner product-based tiling requires fetching an entire row (or column) for each output element, leading to an arithmetic intensity of only $\frac{k}{k} = 1$. In contrast, the outer product formulation enables multiple accumulations over a tile, where fetching m elements from A and n elements from B can yield mn multiply-accumulate (MAC) operations, resulting in a potentially much higher arithmetic intensity: $\frac{mn}{m+n}$. This higher intensity translates directly to greater computational efficiency, particularly on memory-bandwidth-limited architectures.

Rosko builds upon this outer product approach, combining it with smart scheduling (such as CAKE) and packing, to not only maximize arithmetic intensity but also efficiently exploit sparsity patterns typically arising in pruned deep neural network models.

4.3 Related Works

We now position Rosko in relation to prior work on structured sparsity for hardware efficiency and zero-skipping computation, highlighting its unique contributions.

4.3.1 Achieving Fine-Grained Structured Sparsity

As sparsity increases, there is a tradeoff between enabling granular sparsity and maintaining efficient hardware support. Prior works such as [25] employ pruning techniques that ensure dense, contiguous groups of non-zero weights, sized to and aligned to the width of SIMD hardware (e.g., vector registers). While such pruning enables efficient use of SIMD hardware, it limits how granular sparsity can be while maintaining high model accuracy.

Balanced sparsity [23], also known as N:M sparsity, aligns weights to groups of size M; each group contains N nonzero weights with $N < M$. Approaches such as [17] specifically target 2:4 sparsity ($N = 2, M = 4$) to allow specialized hardware support (e.g., Nvidia A100's Sparse Tensor Cores [5]).

With Rosko, we obtain a matrix with a target sparsity (e.g., 90%) and sparse columns through iterative pruning during training. Since the pruned elements are from column input vectors, we may vary the size of the columns without restricting our pruning granularity to hardware SIMD lengths (Sect. 4.6) or specific sparsities (e.g., 50% in 2:4 sparsity).

4.3.2 Mechanisms for Zero Skipping

Prior works analogous to Rosko's row skipping include zero-gating and zero-skipping methods. Zero-gating hardware, such as in [3, 24], disables computation upon loading a zero operand; this conserves energy, but does not directly increase computational throughput. In contrast, zero-skipping approaches such as [4, 10] avoid loading or computing with zero operands altogether, often by using compressed storage formats such as CSR [7], CSC [6], COO, or CSR5 [16]. Although these formats prevent wasted computation and storage, they can introduce metadata overhead and irregular memory accesses, which can limit practical hardware acceleration [18].

Further, nonzeros may be unevenly distributed, leading to workload imbalances across compute units. Specialized hardware and input formats, as in [1], alleviate these issues by adopting restrictive sparsity patterns such as 2:4 sparsity; other works [19] add routing logic for dynamic workload balancing.

In Rosko, outer product column sparsity enables efficient zero-skipping: computations corresponding to zeros in a column input vector are skipped directly in the outer product step, and dense row vectors corresponding to empty columns are not loaded at all. Notably, Rosko's scheme does not require hardware modifications, and can be implemented on existing CPUs while still benefiting from SIMD vectorization.

In summary, Rosko occupies a previously under-served regime, offering fine-grained, hardware-friendly structured sparsity and efficient computation skipping without the overheads or limitations of prior techniques.

4.4 Outer Product Row Skipping: Overview, Motivations, and Innovation

In this section, we review the motivations for Rosko's proposed outer product row skipping approach, summarize its innovations, and provide an overview.

4.4.1 Motivation and Overview

Outer product-based matrix multiplication (MM) allows for efficient computations by accumulating outer products to compute $C = A \times B$. In Rosko, we exploit this structure to skip computations associated with zero-valued elements in the column vectors of A, resulting in what we call *row skipping outer products*. Figures 4.2 and 4.3 illustrate this computation, where A is the weight matrix, B is the data matrix, and C is the accumulation target.

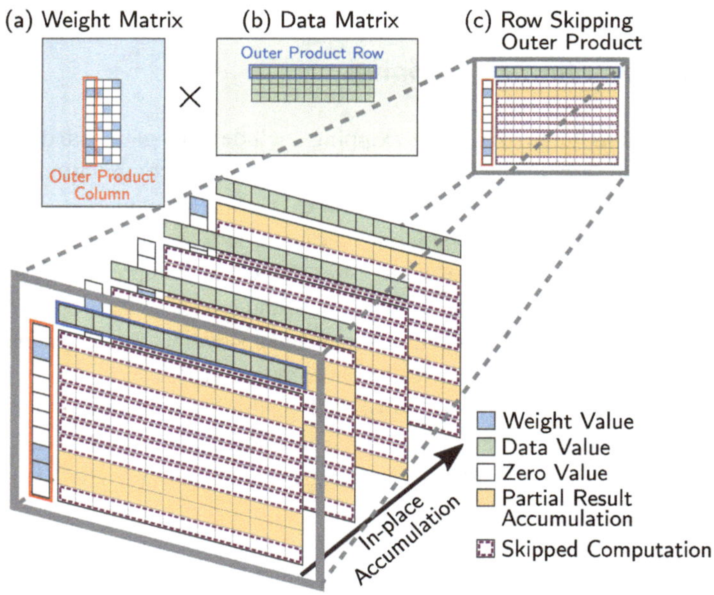

Fig. 4.2 Rosko overview and illustration of outer product MM. (**a**) Sparse weight matrix columns are multiplied with (**b**) dense data matrix rows using (**c**) row skipping outer products. Multiple outer products accumulate in-place during an MM

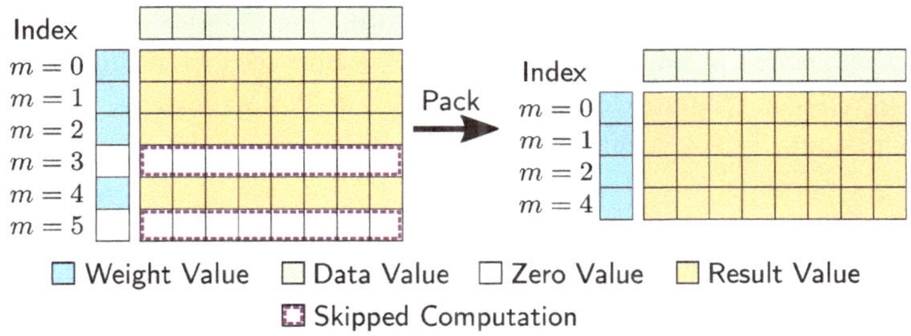

Fig. 4.3 Example of a row skipping outer product between a sparse column (blue) and dense row (green) to obtain a result matrix (yellow). White indicates zero values, and purple outlines indicate skipped rows of computation. Here, row computations for indices $m = 3$ and $m = 5$ are skipped and omitted from the resulting packed computation. A higher column sparsity results in more skipped rows

Each outer product in this formulation is between a column vector a of A and a corresponding row vector b of B. Each row in the outer product depends only on a single element from a, so it is possible to skip a complete row of computation simply by checking whether that element is zero. When an element has been skipped for a given outer product, it will not participate in any further computations, enabling an efficient streaming scheme in which only nonzero elements of a are loaded and multiplied with stationary elements of b.

By leveraging this streaming scheme, Rosko enables outer product column sparsity: computations and memory accesses for zero-valued column elements can be entirely omitted, as can loading the corresponding dense row vectors when a whole input column is empty. The sparse column inputs can then be efficiently packed, with low overhead, into dense storage suitable for SIMD execution (Fig. 4.3). This approach is simpler and more efficient than conventional inner product approaches, which require constant checking and rarely produce coalesced workloads for vector hardware.

4.4.2 Summary of Contributions

A notable advantage of Rosko's row skipping outer product is that, unlike some prior approaches [25], it is straightforward to implement on general-purpose SIMD architectures and does not require hardware modification. This approach maintains high performance at high sparsity, even for non-contiguous nonzero weights. The necessary indexing scheme for the column-packed inputs is simple, minimizing overhead. To summarize, the core innovations of Rosko's approach are:

- **Outer product column sparsity enables row skipping:** By pruning for sparsity in column input vectors, Rosko can skip entire rows of computation based on the presence of zero elements, dramatically reducing required operations.

- **Fine-grained and accurate structured pruning:** We demonstrate it is sufficient to perform fine-grained pruning on column input vectors to achieve high model sparsity without significant accuracy loss. For example, Rosko achieves 97% model sparsity while maintaining 75.8% accuracy on CIFAR-100 (Sect. 4.7.4).

These innovations allow Rosko to provide efficient, hardware-friendly, and highly sparse matrix multiplications, bridging the performance gap for sparsity regimes that are poorly served by conventional libraries.

4.5 Outer Product Column Sparsity

Outer product column sparsity capitalizes on sparsity present in the input columns used for outer product computations. This idea is broadly applicable to any operation reducible to matrix multiplications, including standard convolutions and transformer layers. Unlike traditional structured pruning approaches, which are based on network constructs such as filters, channels, or layers [2], outer product column sparsity directly targets the computational structure of the matrix operation.

As illustrated in Fig. 4.4a, Rosko pruning targets sparsity within the outer product columns of a weight matrix—for example, one derived from a convolutional layer—enabling row skipping in subsequent computations. This approach naturally supports simple and effective pruning for load-balanced computations: by pruning each outer product column to the same sparsity, each computation block receives a similar amount of work, which is important for maximizing utilization on parallel hardware.

Empirical results in Fig. 4.4b demonstrate that Rosko pruning achieves high model sparsity while maintaining accuracy on real tasks. Specifically, on ResNet-50 and CIFAR-100, Rosko pruning enables efficient row skipping and delivers test accuracy comparable to unstructured pruning. Additional results for other networks and datasets can be found in Fig. 4.5.

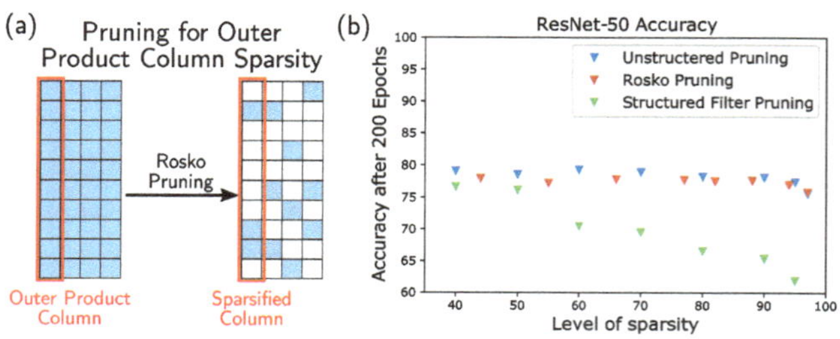

Fig. 4.4 (**a**) Example of Rosko pruning of outer product columns to obtain the weight matrix of Fig. 4.2. (**b**) CIFAR-100 test accuracy after 200 epochs on ResNet-50 with Rosko pruning. Rosko enables row skipping during computation while achieving a model accuracy comparable to unstructured pruning. See Fig. 4.5 for results on additional networks

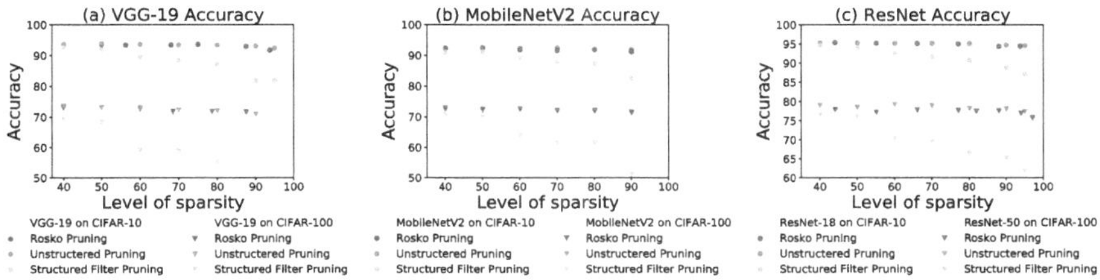

Fig. 4.5 Rosko pruning shows little to no loss in model accuracy compared to unstructured pruning after 200 epochs. For sparsities greater than 50%, Rosko pruning maintains higher accuracy compared to structured pruning using 3×3 blocks. (**a**) Rosko pruning for VGG-19, (**b**) Rosko Pruning for MobileNetV2, and (**c**) Rosko pruning for ResNet-18 and ResNet-50

An important advantage is the facilitation of load-balanced computations: by pruning each outer product column to the same sparsity, each computation block receives an equal share of work. Rosko achieves near-ideal speedup across a wide range of sparsities on GPUs, demonstrating this load balancing in practice. While this echoes group-based approaches (such as balanced or $N{:}M$ sparsity) [23, 26], Rosko does not require enforcing identical sparsity across all outer product groups to benefit from row skipping.

4.6 Training for Row Skipping

To realize the benefits of row skipping, we employ an iterative pruning strategy integrated into model training. Starting from a dense model, we iteratively prune weights with the smallest magnitudes every 10 epochs until the target sparsity is reached, subsequently continuing training until a total of 200 epochs have elapsed. This process is applied to Rosko pruning, unstructured pruning, and structured filter pruning.

For Rosko, pruning is performed within each outer product column, promoting uniform sparsity across columns and ensuring even skipping of rows during computation. This aligns with $N{:}M$ sparsity in spirit—selecting a fraction of weights for removal within a group—but, importantly, groups are defined over outer product columns rather than arbitrary or row-wise groupings (Fig. 4.4a). This distinction facilitates optimal computation skipping and load balancing. In contrast, $N{:}M$ pruning focused on inner products would produce row-wise weight pruning.

Our experiments on standard classification networks—VGG-19 [20], MobileNetV2 [12], ResNet-18, and ResNet-50 [11]—using CIFAR-10 and CIFAR-100 [15] datasets (Fig. 4.5) demonstrate that Rosko pruning attains high sparsity without compromising model accuracy. For example, Rosko achieves 97% model sparsity with 75.8% accuracy on ResNet-50 for CIFAR-100, matching the accuracy of unstructured pruning and outperforming structured filter pruning. Overall, this approach enables efficient, hardware-friendly pruning that maintains accuracy while unlocking substantial computational savings.

4.7 Performance Evaluation of Rosko on CPUs

4.7.1 CPU Evaluation Setup

We implemented Rosko in C++ using Intel AVX2 and ARM Neon SIMD intrinsics, with CAKE (Chap. 2) scheduling for maximal parallelism and memory efficiency. Rosko is evaluated on two distinct architectures: the Intel Core i9-10900K (high-performance desktop, 41 GB/sec DRAM) and ARM Cortex-A72 (mobile, 4.4 GB/sec DRAM), enabling analysis under different bandwidth and memory constraints.

4.7.2 Evaluation Methodology

We benchmark Rosko on both CPUs against leading dense and sparse matrix multiplication libraries. GFLOPs/sec is the metric for sparse matrix multiplication, while runtime is used for dense matrix multiplication comparisons due to differing operation counts. DRAM bandwidth and I/O are also reported. Sparse matrices evaluated span machine learning inference tasks with a range of sparsity (75%–98%).

4.7.3 Ablation Study

To quantify contributions of Rosko's optimizations, we ablate key components (row skipping, computation shaping, reordering, and packing). As shown in Fig. 4.6, each technique improves DRAM efficiency, throughput, and branch prediction, with all together delivering the best performance. Notably, column density reordering reduces CPU misspeculation and bandwidth demands.

4.7.4 DNN Inference Latency

For networks pruned for outer product column sparsity (e.g., ResNet-50 on CIFAR-100), Rosko achieves lower end-to-end inference latency compared to CAKE, ARMPL, and ARMCL, especially as sparsity increases (Fig. 4.7).

4.7.5 Extreme Sparsity (>99%)

Rosko is also evaluated against oneMKL on highly sparse ML-graph matrices from SuiteSparse [14]. On Intel CPUs and all tested matrices (99.87–99.97% sparsity), Rosko delivers up to 4.3× higher throughput while consuming less bandwidth (Fig. 4.8), outperforming especially when sparsity is irregular or random.

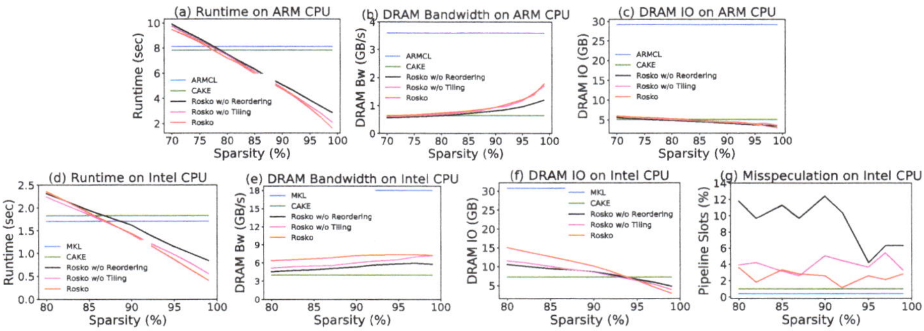

Fig. 4.6 Rosko performance results with various features ablated. We run sparse matrix multiplication between 5000 × 5000 matrices on ARM and 10,000 × 10,000 matrices on Intel, measuring DRAM bandwidth usage, runtime, and CPU misspeculation. Rosko uses CAKE scheduling to reduce DRAM bandwidth usage and significantly reduce runtime relative to ARMCL ((**a**) and (**b**)) and MKL ((**d**) and (**e**)). In (**c**) and (**f**), Rosko can reduce DRAM I/O beyond CAKE as sparsity increases. (**a**) and (**d**) show sparsity-aware tiling can further improve runtime as sparsity increases. However, at high sparsities, Rosko's DRAM bandwidth usage is slightly higher than CAKE. Computation and memory access skipping increases the need to fetch new data into each core for subsequent computation. On the Intel CPU, the irregular distribution of zeros causes up to 13% of pipeline slots to be wasted due to misspeculation (**g**). Rosko's reorders columns by density, allowing the CPU to better predict outer product loop bounds

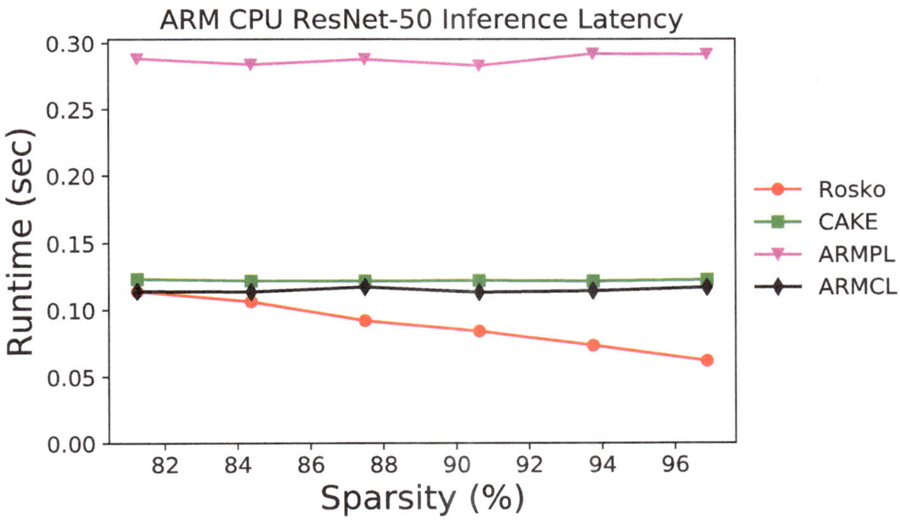

Fig. 4.7 ARM Cortex-A72 inference latency on ResNet-50 models trained on the CIFAR-100 dataset for various levels of outer product column sparsity (denoted as %). We use all 4 cores on the Cortex-A72

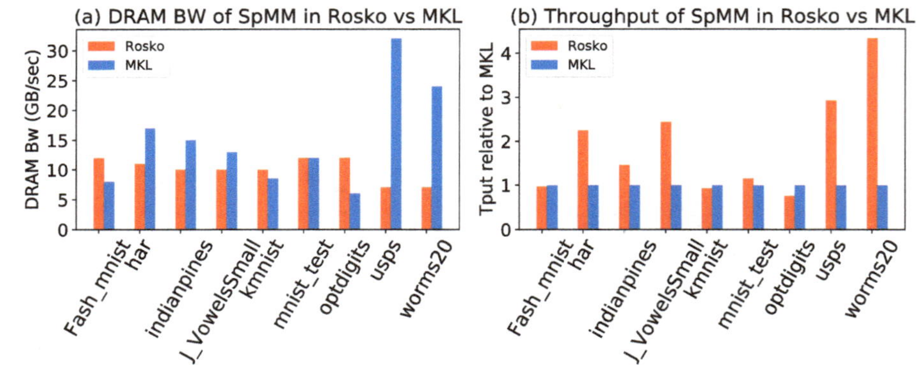

Fig. 4.8 Performance of Rosko and oneMKL on SpMM problems from the ML_Graph group of the SuiteSparse benchmark. Selected matrices have sparsities between 99.87 and 99.97%, and fit into DRAM memory. These results show that Rosko performs well on these highly sparse matrices

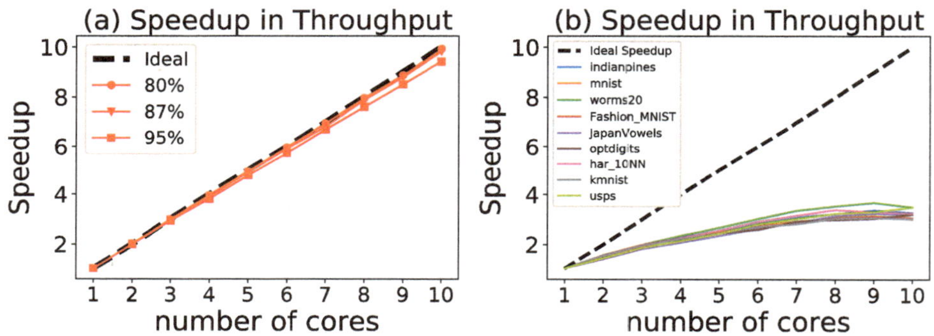

Fig. 4.9 Load balancing across CPU cores for SpMM with outer product column sparsity (**a**) and with unstructured random sparsity (**b**). In (**a**), pruning ensures each computation block contains the same number of nonzeros, achieving near-ideal scaling for 10 cores at various sparsity levels. In (**b**), unstructured random sparsity typical of graph-ML applications results in imbalanced nonzero distributions, reducing multi-core efficiency

4.7.6 Load Balancing Across Cores

Pruning outer product columns to uniform sparsity achieves near-ideal multi-core scaling, as shown in Fig. 4.9a. Load balance degrades with unstructured random sparsity (Fig. 4.9b).

4.8 Summary

This chapter introduced Rosko, a row-skipping algorithm that exploits outer product column sparsity for efficient sparse matrix multiplication. By skipping entire rows of computation corresponding to zero entries in input columns, Rosko achieves significant reductions in unnecessary operations with minimal sparsity management overhead. Its simple indexing scheme enables flexible and fine-grained pruning while preserving high model accuracy across a wide range of sparsity levels.

Rosko is well-suited for SIMD architectures and achieves strong load balancing, integrating seamlessly with efficient computation schedules that maximize data reuse and minimize IO. Empirical results show that Rosko consistently outperforms state-of-the-art matrix multiplication libraries across diverse sparsity regimes.

Overall, Rosko presents a practical approach to efficient computation in sparse neural networks, advancing both the theory and engineering of sparse deep learning inference on modern hardware.

References

1. NVIDIA Ampere Architecture In-Depth. https://developer.nvidia.com/blog/nvidia-ampere-architecture-in-depth/, May 2020.
2. Sajid Anwar, Kyuyeon Hwang, and Wonyong Sung. Structured Pruning of Deep Convolutional Neural Networks. *ACM Journal on Emerging Technologies in Computing Systems*, 13(3):32:1–32:18, February 2017.
3. Yu-Hsin Chen, Joel Emer, and Vivienne Sze. Eyeriss: A Spatial Architecture for Energy-Efficient Dataflow for Convolutional Neural Networks. In *2016 ACM/IEEE 43rd Annual International Symposium on Computer Architecture (ISCA)*, pages 367–379, June 2016.
4. Yu-Hsin Chen, Tien-Ju Yang, Joel Emer, and Vivienne Sze. Eyeriss v2: A Flexible Accelerator for Emerging Deep Neural Networks on Mobile Devices. *IEEE Journal on Emerging and Selected Topics in Circuits and Systems*, 9(2):292–308, June 2019.
5. NVIDIA Corporation. Accelerating inference with sparsity using the NVIDIA ampere architecture and NVIDIA tensorRT. https://developer.nvidia.com/blog/accelerating-inference-with-sparsity-using-ampere-and-tensorrt/, 2021.
6. Jack Dongarra. Compressed Column Storage (CCS). http://netlib.org/linalg/html_templates/node92.html.
7. Jack Dongarra. Compressed Row Storage (CRS). http://netlib.org/linalg/html_templates/node91.html.
8. Trevor Gale, Matei Zaharia, Cliff Young, and Erich Elsen. Sparse GPU kernels for deep learning. In *Proceedings of the International Conference for High Performance Computing, Networking, Storage and Analysis*, SC '20, pages 1–14, Atlanta, Georgia, November 2020. IEEE Press.
9. Kazushige Goto and Robert A. van de Geijn. Anatomy of high-performance matrix multiplication. *ACM Trans. Math. Softw.*, 2008.
10. Song Han, Xingyu Liu, Huizi Mao, Jing Pu, Ardavan Pedram, Mark A. Horowitz, and William J. Dally. EIE: Efficient inference engine on compressed deep neural network. *ACM SIGARCH Computer Architecture News*, 44(3):243–254, June 2016.
11. Kaiming He, Xiangyu Zhang, Shaoqing Ren, and Jian Sun. Deep residual learning for image recognition, 2015.
12. Andrew G. Howard, Menglong Zhu, Bo Chen, Dmitry Kalenichenko, Weijun Wang, Tobias Weyand, Marco Andreetto, and Hartwig Adam. Mobilenets: Efficient convolutional neural networks for mobile vision applications, 2017.
13. Intel Corporation. *Intel oneAPI Deep Neural Network Library*, March 2016.

14. Scott P Kolodziej, Mohsen Aznaveh, Matthew Bullock, Jarrett David, Timothy A Davis, Matthew Henderson, Yifan Hu, and Read Sandstrom. The suitesparse matrix collection website interface. *Journal of Open Source Software*, 4(35):1244, 2019.
15. Alex Krizhevsky. CIFAR-10 and CIFAR-100 datasets. https://www.cs.toronto.edu/~/cifar.html.
16. Weifeng Liu and Brian Vinter. CSR5: An Efficient Storage Format for Cross-Platform Sparse Matrix-Vector Multiplication. In *Proceedings of the 29th ACM on International Conference on Supercomputing*, ICS '15, pages 339–350, New York, NY, USA, June 2015. Association for Computing Machinery.
17. Asit Mishra, Jorge Albericio Latorre, Jeff Pool, Darko Stosic, Dusan Stosic, Ganesh Venkatesh, Chong Yu, and Paulius Micikevicius. Accelerating Sparse Deep Neural Networks. *arXiv:2104.08378 [cs]*, April 2021.
18. Sharan Narang, Eric Undersander, and Gregory Diamos. Block-Sparse Recurrent Neural Networks. *arXiv:1711.02782 [cs, stat]*, November 2017.
19. Eric Qin, Ananda Samajdar, Hyoukjun Kwon, Vineet Nadella, Sudarshan Srinivasan, Dipankar Das, Bharat Kaul, and Tushar Krishna. SIGMA: A Sparse and Irregular GEMM Accelerator with Flexible Interconnects for DNN Training. In *2020 IEEE International Symposium on High Performance Computer Architecture (HPCA)*, pages 58–70, February 2020.
20. Karen Simonyan and Andrew Zisserman. Very deep convolutional networks for large-scale image recognition, 2015.
21. Field G. Van Zee and Robert A. van de Geijn. BLIS: A framework for rapidly instantiating BLAS functionality. *ACM Transactions on Mathematical Software*, 41(3):14:1–14:33, June 2015.
22. Zhang Xianyi, Wang Qian, and Zaheer Chothia. Openblas. *URL:* http://xianyi.github.io/OpenBLAS, page 88, 2012.
23. Zhuliang Yao, Shijie Cao, Wencong Xiao, Chen Zhang, and Lanshun Nie. Balanced Sparsity for Efficient DNN Inference on GPU. In *AAAI Conference on Artificial Intelligence (AAAI)*, January 2019.
24. Lin Ye, Jinghao Ye, Masao Yanagisawa, and Youhua Shi. A Zero-Gating Processing Element Design for Low-Power Deep Convolutional Neural Networks. In *2019 IEEE Asia Pacific Conference on Circuits and Systems (APCCAS)*, pages 317–320, November 2019.
25. Jiecao Yu, Andrew Lukefahr, David Palframan, Ganesh Dasika, Reetuparna Das, and Scott Mahlke. Scalpel: Customizing DNN pruning to the underlying hardware parallelism. In *2017 ACM/IEEE 44th Annual International Symposium on Computer Architecture (ISCA)*, pages 548–560, June 2017.
26. Aojun Zhou, Yukun Ma, Junnan Zhu, Jianbo Liu, Zhijie Zhang, Kun Yuan, Wenxiu Sun, and Hongsheng Li. Learning N:M Fine-grained Structured Sparse Neural Networks From Scratch. *arXiv:2102.04010 [cs]*, April 2021.

Efficient Neural Network Computation with Rank-Sliced Gather-Scatter Activation

5

Chapter Overview

The previous chapters explored optimizing matrix multiplication by careful hardware-aware tiling and leveraging sparsity. However, tiling and sparsity optimization often treat the underlying data as immutable. In this chapter, we investigate how compressing the matrix itself can yield further savings in both memory and computation. Specifically, we focus on using Singular Value Decomposition (SVD) to represent weight matrices in a more compact form of neural network layers.

Once a matrix is stored in its SVD form, the challenge becomes efficiently computing activations: How can we multiply incoming inputs by SVD-compressed weights without incurring the overhead of reconstructing the original dense matrix or performing multiple inefficient matrix multiplications? To address this, we propose a method that directly leverages the structure of the SVD representation. Our approach uses a gather-scatter strategy that processes each rank in the decomposition independently, allowing us to realize memory savings and computational efficiency.

By rethinking how we operate on compressed matrices, we show how it is possible to save memory bandwidth and reduce computation.

5.1 Introduction

The efficient computation of neural network activations is a core challenge as models increase in size and complexity. Many modern architectures, such as transformers and deep convolutional networks, rely heavily on large matrix multiplications, which can create significant bottlenecks in terms of both computation and memory I/O. An effective strategy to address these challenges is to represent weight matrices in factored or low-rank forms,

A. Sabot, *Minimizing Data Movement and Parameter Count Across the Machine Learning Stack*, Synthesis Lectures on Computer Science,
https://doi.org/10.1007/978-3-032-23100-0_5

such as the singular value decomposition (SVD) $W = U\Sigma V^T$. However, realizing the full I/O and computational benefits of these structured representations requires algorithms that are specifically designed to exploit rank-sliced storage and inherent sparsity.

This chapter presents the **Gather-Scatter Activation (GASA)**[1] algorithm, a novel method for efficiently computing neural network activations when layer weights are stored in SVD form. GASA operates in a *rank-sliced* fashion: each slice from the SVD of W is treated as a rank-1 matrix $S = uv$, where u and v are a column and a row vector, respectively. The complete activation XW is calculated as a sum over these slices.

At the heart of GASA is the gather-scatter approach. For each slice, the algorithm proceeds in two stages: (1) *Gather*, where columns of the data matrix X are multiplied and combined according to the pattern in the vector u, producing a gathering vector x_g; and (2) *Scatter*, in which this vector is distributed to columns of XS according to v. By repeating this process for each rank-1 slice and accumulating all the results, the full activation XW is efficiently computed. This process is illustrated in Figs. 5.1 and 5.2.

This approach is not only hardware-friendly and conducive to SIMD/vectorized execution, but also well suited for leveraging sparsity often present in the SVD factors of pruned or compressed networks. Furthermore, GASA achieves provably optimal I/O complexity under practical assumptions, making it highly suitable for deployment on memory- and bandwidth-constrained systems.

The remainder of this chapter details the GASA algorithm, provides theoretical and architectural analysis, and demonstrates its practical efficiency in neural network activation computation. The next chapter will focus on methods for generating and maintaining sparsity in SVD-based models such that activations, when computed with GASA, are both efficient and accurate.

5.2 Background and Motivation

Matrix-matrix multiplication is a core operation in deep learning, underlying the forward and backward passes of nearly every neural network layer. As models scale, their weight matrices often become too large to store or process efficiently in their raw dense form. This has motivated the use of structured representations such as low-rank factorizations.

The singular value decomposition (SVD) is one such structure, expressing any matrix $W \in \mathbb{R}^{n\times n}$ as $W = U\Sigma V^T$, where U and V are orthogonal matrices and Σ is diagonal [1,4]. When W is low-rank or can be well-approximated by a low-rank matrix, storing only the leading r singular vectors/values provides substantial gains in storage and, potentially, computation.

Traditional implementations, however, often do not fully exploit the opportunities for memory-access and computation minimization inherent to SVD-form weights. Standard

[1] The name GASA stands for **Gather–Scatter Activation**.

matrix multiplication algorithms require frequent moves of large data between memory hierarchies, and this is a major bottleneck, especially on modern memory-bound systems. The gather-scatter paradigm introduced here seeks to address these limitations. Additionally, the GASA method can leverage sparsity which is commonly seen in deep learning workloads.

5.2.1 Gather-Scatter

Gather-scatter algorithms are widely used in fields such as signal processing, scientific computing, and parallel computing. They operate in two primary stages: a gather phase, where data is read from external memory, and a scatter phase, where the computed data is written back to the memory system or an output matrix [2, 3]. In this work, we extend the gather-scatter approach to a new area, that is, efficient computation of activations between a data matrix X and a weight matrix represented in SVD form.

5.3 The Gather-Scatter Algorithm

5.3.1 Rank-Sliced Decomposition

Consider $W = U\Sigma V^T$ with reduced rank r; then $W = \sum_{i=1}^{r} \sigma_i u_i v_i^T$, where u_i and v_i are the ith columns of U and V, and σ_i is the ith diagonal element of Σ. Each term $\sigma_i u_i v_i^T$ is a rank-1 "slice" S_i of W.

The key idea of GASA is to compute XS_i for each slice individually and then sum the results, weighted by σ_i, to obtain XW:

$$XW = \sum_{i=1}^{r} \sigma_i X S_i.$$

Figure 5.1a depicts a rank-1 slice of W.

5.3.2 Gather and Scatter Stages

For a single slice $S = uv$, the multiplication XS can be efficiently computed in two stages:

1. **Gather** stage: Combine the columns of X according to the pattern in u, producing the vector $x_g = Xu$.
2. **Scatter** stage: Distribute x_g across the columns of the output according to v, so $XS = x_g v$.

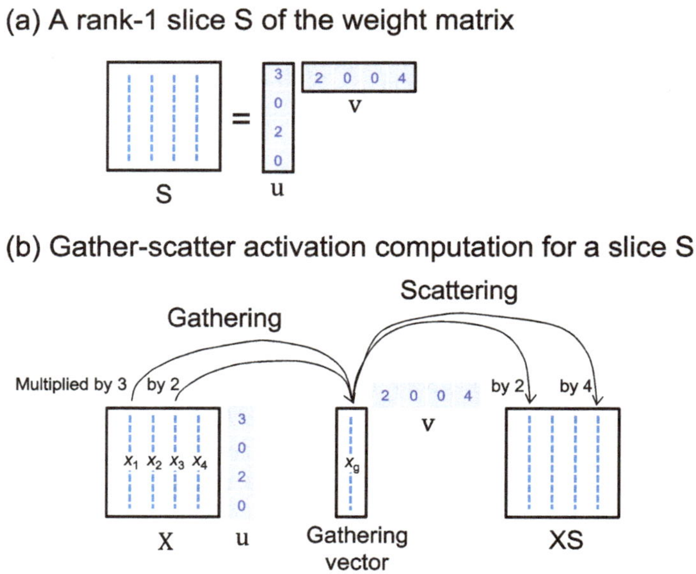

Fig. 5.1 The **gather-scatter algorithm** of computing activations for a rank-1 slice S of a weight matrix W. (**a**) A rank-1 slice S of W is the product of a column vector u and a row vector v. (**b**) For an activation data matrix X, we compute XS for a slice S in two stages. In the first stage, we gather column vectors of X to form a gathering vector x_g, according to u. In the second scattering stage, we distribute x_g to form columns of the result matrix XS, according to v

With this structure, the computation for a single rank-1 slice only requires linear combinations of columns at each stage—a pattern that is highly amenable to both memory-efficient implementation and exploitation of input/output sparsity.

Figure 5.1 depicts this two-stage process for a rank-1 slice.

5.4 I/O and Computational Complexity Analysis

5.4.1 Working Set and Local Memory

Suppose X and W are both $n \times n$ matrices and W is of rank r. We compute activations XW by computing XS_1, XS_2, XS_3 for all rank slices S_i and accumulating the result $\sigma_1 XS_1 + \sigma_2 XS_2 + \sigma_3 XS_3$. The gather-scatter algorithm requires a working set consisting of r gathering vectors plus intermediate result vectors, all of size n, which must fit in fast (local) memory, as depicted in Fig. 5.2b.

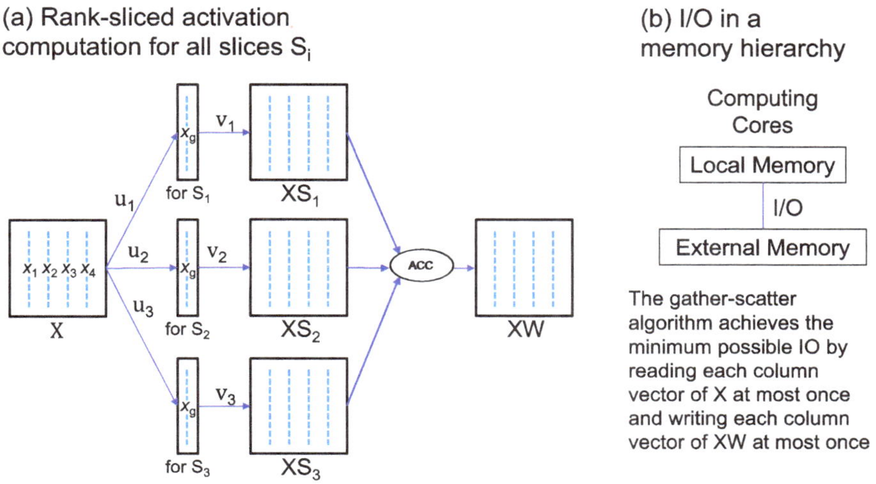

Fig. 5.2 (**a**) We compute activations XW, given a data matrix X and weight matrix W, by computing XS_i for all slices S_i of W. Here, W consists of three slices: S_1, S_2 and S_3, where $S_i = u_i v_i$. We first compute XS_1, XS_2, and XS_3, and then perform weighted accumulation of these results with σ_1, σ_2 and σ_3 being the weighting coefficients, as denoted in the ACC circle in the diagram. (**b**) We can minimize I/O on a memory hierarchy. Suppose that X, as well as u_i and v_i, for all slices, are initially stored in the external memory and that computed results XW are to be written to the external memory. Suppose further that the local memory can hold the gathering vectors x_g for all slices plus a few intermediate vectors to support data streaming. Then by scheduling the gather-scatter algorithm achieves the minimum possible I/O

5.4.2 I/O Efficiency

The computation can minimize I/O as described below. When computing the 3 gathering vectors for XS_1, XS_2, and XS_3 in the gathering stage, we will use local memory to cache the intermediate vectors. After a column vector of X is read from the external memory, scaled copies of this column vector are added to intermediate gathering vectors that correspond to the destination gathering vectors required for this column vector. This implies that we only need to read each column vector of X from the external memory at most once. Therefore, a total of at most n vector reads are required for the gathering operations for all XS_i's.

5.4.3 Operation Counts and Sparsity

Let α denote the fraction of zero entries (sparsity) in u and v. For each slice, both the gather and scatter operations can skip computations at zero locations, yielding a total of $2(1-\alpha)n^2$ multiplications per slice (plus a similar number of additions). The benefits are compounded for highly sparse slices and for multiple slices.

5.4.4 Comparison to Standard MM

Standard (dense) matrix multiplication under the same local memory constraints requires $\Omega(n^2)$ vector reads and often cannot exploit sparsity patterns as directly or efficiently as the gather-scatter scheme.

5.5 Implementation Considerations

5.5.1 Data Layout and Memory Strategies

For maximal efficiency, X and W should be stored in column-major order if possible, since the gather stage processes columns sequentially. The u and v vectors for all slices are ideally buffered in local memory. If additional ranks or slices are needed, multiple passes can be scheduled, or slices can be computed in blocks.

5.5.2 Exploiting Sparsity

Efficient implementation takes advantage of sparse representations for u and v: only nonzero entries are processed, and indexing arrays can be used to skip zero-valued computations and memory accesses.

5.5.3 Parallel and SIMD Execution

The gather and scatter stages are parallelizable. SIMD instructions can vectorize both the aggregation in the gather stage and the distribution in the scatter stage. When multiple slices are needed, they can be processed in parallel, subject to memory bandwidth limits.

5.6 Discussion

The GASA algorithm is broadly applicable to layers in convolutional and transformer architectures wherever SVD-structured weights are available. Its I/O optimality makes it especially relevant for memory-bound hardware accelerators and edge devices. Additionally, the gather-scatter pattern is flexible and can be composed with batching, hybrid sparsity, and mixed-precision techniques.

Limitations include potential overhead from managing many ranks in high-rank matrices and sensitivity to local memory size. Future work may address batching strategies, mixed DNN operations, and integration with distributed training.

5.7 Summary

This chapter introduced the Gather-Scatter Activation (GASA) algorithm, an efficient, sparsity-exploiting, and I/O-optimal method for computing neural network activations when weights are stored in SVD form. By decomposing computation into rank slices and leveraging input/output sparsity patterns, GASA enables significant reductions in data movement and overall computational cost. These strengths lay the foundation for even greater efficiency when combined with structured sparsification and adaptive fine-tuning techniques, which I discuss in the following chapter.

Future work in this direction may involve creating a library for running GASA with a bandwidth aware tiling method that could be derived using the computation space introduced in Chap. 2.

References

1. G. H. Golub and C. Reinsch. Singular value decomposition and least squares solutions. *Numer. Math.*, 14(5):403–420, April 1970.
2. Bingsheng He, Naga K. Govindaraju, Qiong Luo, and Burton Smith. Efficient gather and scatter operations on graphics processors. In *SC '07: Proceedings of the 2007 ACM/IEEE Conference on Supercomputing*, pages 1–12, 2007.
3. Peter Pacheco. *An introduction to parallel programming*. Elsevier, 2011.
4. Gilbert Strang. *Linear algebra and learning from data*. SIAM, 2019.

6 Two-Stage Pruning and Sparsity-Preserving Fine-Tuning for SVD-Based Networks

Chapter Overview

Building upon the hardware-aware inference techniques established in Chap. 5, this chapter addresses the critical challenge of model adaptation within the machine learning lifecycle. While compressing a model for static inference is a significant step, real-world deployment often necessitates fine-tuning or task-specific adaptation. The central question we explore in this chapter is: **How can we adapt neural networks maintained in an SVD representation without sacrificing the efficiency gains achieved through compression?**

This chapter presents a two-stage framework that both reduces the size of the model and maintains efficiency during fine-tuning. In the first stage, the weight matrices of the original network are decomposed using SVD and truncated to a lower rank, for example, reducing a rank 768 matrix to rank 128—yielding a compact model of low rank. This compressed model is then fine-tuned to recover as much predictive performance as possible.

In the second stage, we introduce additional sparsity by pruning the row and column vectors of the SVD decomposition. This enables a further reduction in storage and computation. Crucially, we show how to perform sparsity-preserving fine-tuning. Instead of updating all elements, our method adapts only a carefully selected subset of the already sparse SVD vectors, preserving their sparsity pattern. This approach dramatically reduces compute and memory requirements during model adaptation. Comparable parameter-efficient fine-tuning methods like LoRA [6] do not have sparsity compatible updates, resulting in fine-tuned models with significantly less sparsity.

This chapter details how one can maintain efficient, compact models throughout both inference and training.

A. Sabot, *Minimizing Data Movement and Parameter Count Across the Machine Learning Stack*, Synthesis Lectures on Computer Science,
https://doi.org/10.1007/978-3-032-23100-0_6

6.1 Introduction

Neural network compression and adaptation are increasingly vital as models scale and are deployed in resource-constrained environments. While low-rank factorizations such as SVD enable efficient representation of large weight matrices, further improvements in memory and computational efficiency require aggressive pruning of these structures. At the same time, it is essential that such pruning techniques preserve the accuracy and adaptability of the network, particularly when applied to modern fine-tuning paradigms. The method presented in this chapter is complementary to the GASA approach presented in Chap. 5.

This chapter presents a comprehensive framework for **two-stage sparsification** of SVD-based neural networks. The first stage applies *rank pruning* to reduce the dimensionality of the weight matrices, removing singular values and their associated subspaces with minimal impact on accuracy. The second stage, termed *UV pruning*, sparsifies the columns u of U and rows v of V within each rank-1 slice, targeting the fine-grained removal of unimportant weight entries.

I further introduce a sparsity-preserving, parameter-efficient fine-tuning scheme that extends the benefits of pruning into the adaptation of pre-trained models. Unlike prior approaches (e.g., LoRA [6]) that can harm inference time efficiency by introducing dense low-rank updates, this method maintains the exact sparsity structure imposed by pruning, thus retaining the computational gains offered by algorithms like GASA (introduced in Chap. 5).

The chapter combines formal algorithms, empirical evaluations, and comparative analysis to showcase the effectiveness of two-stage pruning and sparsity-preserving adaptation. Together, these tools complete the end-to-end story of how to compress, adapt, and efficiently compute in modern SVD-based deep neural networks.

6.2 Background: Parameter-Efficient Fine-Tuning (PEFT)

Parameter-Efficient Fine-Tuning (PEFT) techniques have become central to the adaptation of large pre-trained neural networks across diverse tasks and domains. The main goal of PEFT is to minimize the number of trainable parameters introduced during fine-tuning, enabling memory- and compute-efficient adaptation without modifying the full capacity or increasing inference overhead.

A widely-adopted approach in PEFT is *Low-Rank Adaptation (LoRA)* [6], which augments pre-trained weights with trainable low-rank matrices. By injecting task-specific information via these small updates, LoRA achieves strong performance while keeping the base model weights fixed and ensuring inference efficiency.

Recent advances in this area further integrate sparsity into the adaptation process. *Masked LoRA* [12] introduces masking over the adaptation matrices, learning sparse task-specific updates and thus further reducing adaptation overhead. *SORA* (Sparse Zero-Rank Adaptation) [2] selectively zeros out specific ranks within the LoRA update, balancing performance with efficient adaptation by focusing only on key low-rank directions. *RoseLoRA* [11] directly enforces row- and column-wise sparsity within the LoRA update matrices, creating an overall sparse adaptation. However, the resulting sparsity pattern does not necessarily align with that of the original weight matrix and may reduce end-to-end sparsity at inference.

Other notable extensions include *DoRA* [8], which decouples the magnitude and direction of weight updates and applies low-rank adaptation only to the directional component, and *LoRA-XS* [1], which utilizes an SVD of the original weight matrix to introduce frozen and unfrozen low-rank updates. While LoRA-XS demonstrates the utility of SVD structure, it does not address or preserve sparsity within the SVD factors themselves.

Despite these advances, most PEFT approaches either introduce additional parameter matrices, rely on explicit masking, or do not preserve the inherent sparsity structure of the base model. In contrast, the techniques developed here focus on fine-tuning strategies that maintain existing sparsity patterns throughout the model and adaptation process, ensuring fully efficient sparse inference without the need for extra masks, auxiliary storage, or dense intermediate computations.

6.3 Two-Stage Pruning: Rank and UV Pruning

The proposed approach begins by factorizing each weight matrix W of a pre-trained model using the singular value decomposition (SVD), $W = U\Sigma V^T$ [3,10]. Two forms of pruning are then applied:

- **Rank Pruning:** (Fig. 6.1a) The smallest singular values in Σ are dropped, reducing the effective rank of W. The resulting model is parameter-efficient and amenable to further sparsification.
- **UV Pruning:** (Fig. 6.1c) To increase sparsity, further pruning is applied directly to the u and v vectors forming the columns of U and V, respectively, independent of Σ. This step can be fine-grained and tailored to each slice of the weight matrix.

This two-stage strategy enables compounded sparsity: if $a\%$ of the singular values are removed during rank pruning, and $b\%$ of the entries in U and V are pruned with UV pruning, the compounded sparsity is $1 - (1 - a)(1 - b)$.

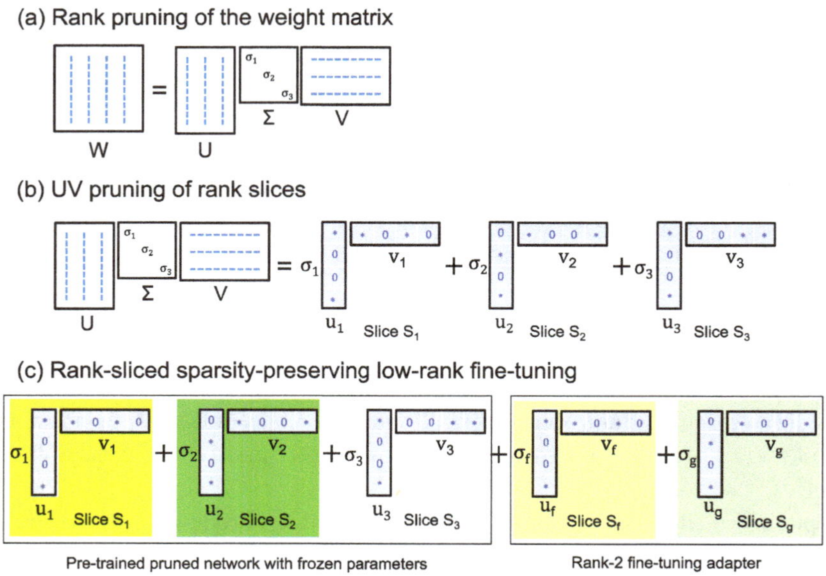

Fig. 6.1 The initial weight matrix W is first decomposed into the SVD representation $W = U\Sigma V^T$. The rank of W is then pruned to create a low-rank representation. Diagram (**a**) depicts an example of a low-rank representation with a rank equal to three and singular values being σ_1, σ_2 and σ_3. Diagram (**b**) shows that $U\Sigma V^T$ is a weighted sum of three rank-1 products $u_1 v_1$, $u_2 v_2$, and $u_3 v_3$ for slices S_1, S_2 and S_3, respectively, where u and v vectors have sparsity $\alpha = 50\%$. Diagram (**c**) illustrates our proposed rank-sliced sparsity-preserving low-rank fine-tuning. The two rank-1 fine-tuning slices S_f and S_g preserve the sparsity structures of slices S_1 and S_2, respectively

6.4 Sparsity-Preserving Low-Rank Fine-Tuning

After two-stage pruning, we fine-tune the pruned model using a *sparsity-preserving low-rank update* strategy. Unlike standard low-rank adaptation methods such as LoRA [6], which introduce dense updates, our approach ensures all updates conform to the original sparsity structure.

In methods like LoRA, the fine-tuning step introduces additional low-rank matrices A and B so that the update to the weight matrix is AB^T. Since A and B are not generally sparse, their product is typically dense, which can destroy the sparsity of the model and increase inference time and memory usage. Furthermore, in LoRA, the parameters of A and B must be stored separately in addition to the original W matrix, doubling the number of parameters required for the adaptation.

By contrast, in our sparsity-preserving approach, we:

- Apply rank-1 updates to a selected subset of weight matrix slices (the ones corresponding to the highest singular values).

- For each updated slice $S = uv$, learn low-rank updates Δu and Δv with the same sparsity patterns as u and v, so that $S + \Delta S$ matches the original sparsity mask.

Because both the model and the adapters share the same sparsity structure, no extra parameter storage is needed for separate dense matrices, and after adaptation, the model remains sparse. This preserves the computational and memory efficiency enabled by sparsity, ensures that inference costs remain low, and allows parameter-efficient adaptation by controlling the number of updated slices (the "adapter rank" analogous to LoRA).

Figure 6.1c illustrates this process: updated slices S_f, S_g are added to their respective original slices, and the overall structure is maintained.

6.5 Experimental Results

We empirically validate this approach on both convolutional and transformer networks, demonstrating that weight matrices are highly amenable to both forms of pruning.

6.5.1 Convolutional Network (ResNet-18) on CIFAR-10

As shown in Fig. 6.2a, even aggressive rank pruning of ResNet-18 [4] on CIFAR-10 [7] yields minimal accuracy loss. Figure 6.2b demonstrates that after up to 85% rank pruning, further pruning of the U matrix can achieve high effective sparsity while still maintaining performance.

6.5.2 Transformer Network (Deberta-v3-Base) on IMDB

Table 6.1 shows results on the Deberta-v3-base [5] model for the IMDB sentiment dataset [9]. Iterative SVD and UV pruning are performed:

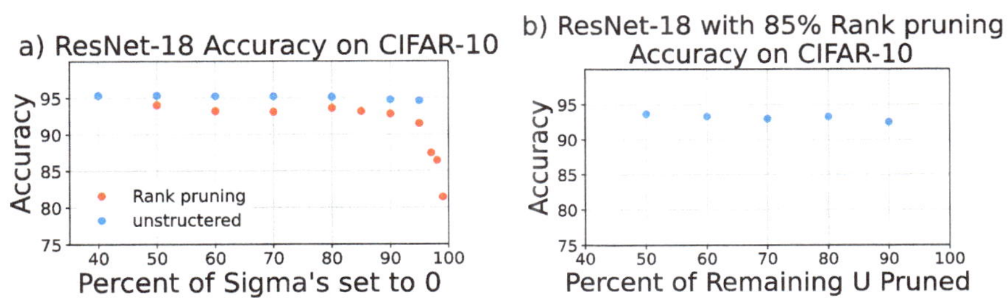

Fig. 6.2 (**a**) Accuracy of ResNet-18 on CIFAR-10 under various levels of rank pruning. (**b**) Further pruning the U matrix after applying 85% rank pruning. The graph is a result of the process in Sect. 6.3 and Fig. 6.1

Table 6.1 Accuracy of Deberta-v3-base [5] on the IMDB dataset [9] under various rank-sliced pruning configurations. Note the initial rank of the model is 768

Rank pruned/rank	No U/V pruning	50% U	50% U and V
0% 768	0.92008	0.88856	0.88592
50% 384	0.90688	0.88892	0.87624
60% 307	0.89652	0.8798	0.8646
70% 230	0.88652	0.86688	0.85052
80% 153	0.8608	0.85268	0.835

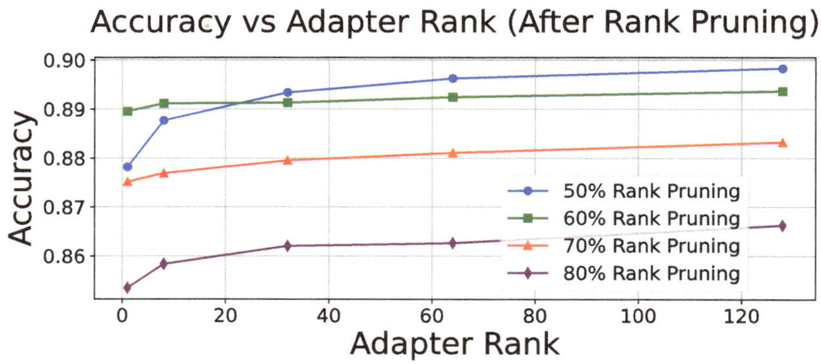

Fig. 6.3 Accuracy of Deberta-v3-base on the IMDB dataset when fine-tuning with a low-rank adapter. Our method, illustrated in Fig. 6.1c achieves close to the performance shown in Table 6.1 while using adapters of small rank, e.g., 20 as opposed to the pre-trained pruned model with rank 150

1. The model is first converted to SVD form and briefly trained.
2. SVD rank is pruned (potentially iteratively) to achieve target sparsity.
3. U and V are pruned further.

As the amount of rank and U/V pruning increases, accuracy declines only modestly. Combining U and V pruning with high-rank pruning results in a compounded parameter reduction with limited model degradation.

6.5.3 Sparsity-Preserving Fine-Tuning with Low-Rank Adapters

We further investigate the use of low-rank adapters for parameter-efficient fine-tuning of pruned models (see Fig. 6.1c). After rank pruning, matrix rank is reduced (e.g., from 768 to 150 in a Deberta-v3-base feedforward layer). During fine-tuning, we freeze the pruned matrix and optimize restricted-rank adapter matrices, which are trained on the target task (IMDB). Results (Fig. 6.3) demonstrate that using adapters of rank as low as 8 can nearly match full-matrix adaptation while updating only a small fraction of the parameters.

6.6 Discussion

The results show that both convolutional and transformer architectures are highly compressible via two-stage SVD-based pruning while maintaining high accuracy. The ability to preserve and leverage sparsity through all stages—including fine-tuning—enables deployment-ready compression that retains both efficiency and adaptability.

A key difference between our approach and LoRA is in parameter storage and inference-time efficiency. LoRA requires maintaining both the frozen original weights and additional A and B dense matrices, which increases parameter count and memory requirements. Moreover, the resulting low-rank adaptation term AB^T is usually dense, removing the benefits of the pruned model's sparsity and increasing inference cost. In contrast, our method applies sparse, rank-sliced low-rank updates directly—preserving both the original sparsity pattern and the lightweight memory/storage footprint—so no extra parameter storage or memory bandwidth is consumed, and inference remains as efficient after fine-tuning as before. This approach integrates seamlessly with the GASA computation algorithm (Chap. 5), maximizing the value of both efficient computation and deep compression.

6.7 Summary

This chapter introduced a two-stage sparsification framework for SVD-structured weights, consisting of rank pruning and UV pruning, and proposed a sparsity-preserving low-rank fine-tuning method. The empirical results demonstrate that these techniques enable substantial parameter and compute reductions in both convolutional and transformer models, while maintaining competitive accuracy and efficient inference.

By avoiding the parameter and inference overheads inherent to methods like LoRA, this approach, together with GASA, establishes a practical path to scalable, adaptive, and efficient deep learning models suitable for modern deployment contexts.

References

1. Klaudia Bałazy, Mohammadreza Banaei, Karl Aberer, and Jacek Tabor. LoRA-XS: Low-rank adaptation with extremely small number of parameters, 2024.
2. Ning Ding, Xingtai Lv, Qiaosen Wang, Yulin Chen, Bowen Zhou, Zhiyuan Liu, and Maosong Sun. Sparse low-rank adaptation of pre-trained language models, 2023.
3. G. H. Golub and C. Reinsch. Singular value decomposition and least squares solutions. *Numer. Math.*, 14(5):403–420, April 1970.
4. Kaiming He, Xiangyu Zhang, Shaoqing Ren, and Jian Sun. Deep residual learning for image recognition, 2015.
5. Pengcheng He, Jianfeng Gao, and Weizhu Chen. DeBERTaV3: Improving DeBERTa using ELECTRA-style pre-training with gradient-disentangled embedding sharing, 2023.

6. Edward J. Hu, Yelong Shen, Phillip Wallis, Zeyuan Allen-Zhu, Yuanzhi Li, Shean Wang, Lu Wang, and Weizhu Chen. LoRA: Low-Rank Adaptation of Large Language Models, October 2021.
7. Alex Krizhevsky. Learning multiple layers of features from tiny images. Technical report, University of Toronto, 2009.
8. Shih-Yang Liu, Chien-Yi Wang, Hongxu Yin, Pavlo Molchanov, Yu-Chiang Frank Wang, Kwang-Ting Cheng, and Min-Hung Chen. Dora: Weight-decomposed low-rank adaptation, 2024.
9. Andrew L. Maas, Raymond E. Daly, Peter T. Pham, Dan Huang, Andrew Y. Ng, and Christopher Potts. Learning word vectors for sentiment analysis. In Dekang Lin, Yuji Matsumoto, and Rada Mihalcea, editors, *Proceedings of the 49th Annual Meeting of the Association for Computational Linguistics: Human Language Technologies*, pages 142–150, Portland, Oregon, USA, June 2011. Association for Computational Linguistics.
10. Gilbert Strang. *Linear algebra and learning from data*. SIAM, 2019.
11. Haoyu Wang, Tianci Liu, Ruirui Li, Monica Cheng, Tuo Zhao, and Jing Gao. RoseLoRA: Row and column-wise sparse low-rank adaptation of pre-trained language model for knowledge editing and fine-tuning, 2024.
12. Junjie Wang, Guangjing Yang, Wentao Chen, Huahui Yi, Xiaohu Wu, Zhouchen Lin, and Qicheng Lao. MLAE: Masked LoRA experts for visual parameter-efficient fine-tuning, 2024.

7 Block-by-Block Knowledge Distillation for Compressed Student Networks

Chapter Overview

Building on the ideas of low-rank approximation and SVD-based compression from Chaps. 5 and 6, this chapter presents an approach to training compact neural networks via block-by-block knowledge distillation. Rather than training a low-rank (SVD-based) model entirely from scratch, we leverage the well-optimized full-rank model to guide the training of its compressed counterpart.

Traditional knowledge distillation techniques typically use the outputs of a full model to supervise the training of a smaller network. In contrast, our method works with a finer granularity by subdividing the models. Neural network architectures are often structured as sequences of repeated blocks, each consisting of one or more layers. We propose distilling knowledge block-by-block, using the activations produced by each block of the full-rank model to directly supervise the corresponding low-rank block in the student model.

This block-wise strategy offers significant advantages. By aligning corresponding blocks between teacher and student, we can substantially boost the performance of low-rank models, making them much more effective than if trained independently. Moreover, this approach facilitates efficient training, as we can focus on training individual blocks or small groups of layers at a time, without requiring backpropagation through the entire model for each update.

This chapter explores the motivation, implementation, and empirical evaluation of block-by-block knowledge distillation, demonstrating how targeted supervision from a high-capacity model can dramatically improve both the accuracy and training efficiency of compressed, low-rank networks.

A. Sabot, *Minimizing Data Movement and Parameter Count Across the Machine Learning Stack*, Synthesis Lectures on Computer Science,
https://doi.org/10.1007/978-3-032-23100-0_7

7.1 Introduction to Building Small Models Using Low-Rank Singular Value Decomposition

The rapid progress of deep learning has yielded models with tremendous performance, but at the cost of ever increasing computational and memory requirements [2, 5, 7, 17]. Vision models such as convolutional neural networks (CNNs) [6, 15] and Vision Transformers (ViTs) [3] excel in image-based tasks, while large language models (LLMs) like GPT [1, 22] lead in natural language processing. However, their sheer size poses deployment challenges in resource-constrained environments and has catalyzed the development of effective model compression techniques.

Knowledge Distillation (KD) [9–13, 24] is a popular approach to model compression, transferring knowledge from a large "teacher" model to a smaller "student" model by training the student to mimic the teacher's outputs and intermediate representations. Most KD methods employ a unified objective, aligning the entirety of the student network with its teacher counterpart. However, as model architectures deepen and become more complex, unified KD objectives exhibit two primary limitations: gradient interference across layers [26] and the accumulation of misalignment errors as the signal propagates through the network.

These issues become especially pronounced in deep or highly modular models, where small discrepancies in earlier layers can propagate forward and impair overall student performance. To address these challenges, we propose **block-by-block knowledge distillation**, a framework which decouples the distillation process for each architectural block—such as Transformer layers in ViTs and LLMs, or bottleneck blocks in ResNets. This design limits cross-block gradient entanglement and contains error accumulation, facilitating more precise student-teacher alignment at every structural stage of the model.

Building on this principle, we introduce a scalable, model-agnostic compression pipeline that begins by constructing a student model directly via *singular value decomposition* (SVD) of the teacher's weight matrices. This low-rank student serves as an efficient architecture-agnostic surrogate, which is then further refined via block-by-block KD to recover—and sometimes even surpass—the performance of its full-rank teacher.

As illustrated in Fig. 7.1, this approach enables substantial compression for both vision and language models—including Swin Transformers and GPT-2—while achieving accuracy comparable to original full-size models. ResNet-50, for example, can be compressed 4× (23.92M down to 5.5M parameters) and still outperform even a deeper but lower-rank baseline like ResNet-18. Similar results hold for large language models and ViTs, where up to 4× compression is achieved with minimal degradation.

Ablation studies in Sect. 7.3.3 validate that isolating knowledge distillation by block outperforms traditional unified KD, confirming the practical value of our framework.

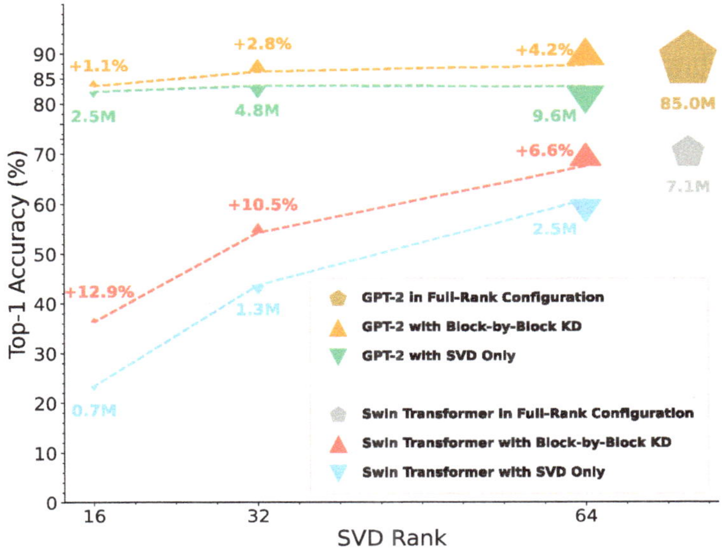

Fig. 7.1 **Low-rank model compression.** Block-by-block KD effectively compresses models while maintaining accuracy comparable to their full-rank counterpart. Performance of GPT-2 (model size excluding word embeddings) on the IMDB dataset and Swin Transformer on the TinyImageNet dataset at ranks 16, 32, and 64. Block-by-block KD improves performance across different compression levels by aligning intermediate representations through isolated, block-wise distillation

Summary of Contributions

1. We introduce a simple, scalable SVD-based construction of student models, directly applying low-rank approximations to the teacher, which is broadly applicable across architectures.
2. We propose block-by-block KD, a framework that reduces gradient interference and error accumulation by isolating the distillation process within individual model blocks.
3. We demonstrate that the combination of SVD compression and block-by-block KD achieves substantial compression and high performance for a range of model families, including CNNs, ViTs, and LLMs.

The remainder of this chapter explores the methodology, empirical validation, and further extensions of the block-by-block knowledge distillation paradigm for low-rank model building.

7.2 Method

7.2.1 Preliminaries

We define a *structural block* (or simply "block") as a computational unit within a model, such as a Transformer block in transformer-based architectures or a bottleneck block in convolutional networks. Each block may include multiple layers (e.g., fully-connected or convolutional) that together produce a unified representation (Fig. 7.2).

Let F_m^{in} and F_m^{out} denote the input and output features of the m-th block in the student model, and G_m^{in}, G_m^{out} the input and output features of the corresponding teacher block. Our method ensures that each block's input and output in the student and teacher models have matching shapes, i.e., $G_m^{\text{in}} = F_m^{\text{in}}$ and $G_m^{\text{out}} = F_m^{\text{out}}$.

We use $N \times C$ to represent feature shapes, where C is always the feature/channel dimension. For CNNs, N represents the spatial size (height $\times$ width), and for transformers, N is the sequence length.

7.2.2 Low-Rank Model Compression with SVD

To create a compact student model, we compress each teacher weight matrix using Singular Value Decomposition (SVD). For each fully-connected or convolutional layer weight matrix $\mathcal{W}$, SVD yields

$$\mathcal{W} = U \Sigma V^T \tag{7.1}$$

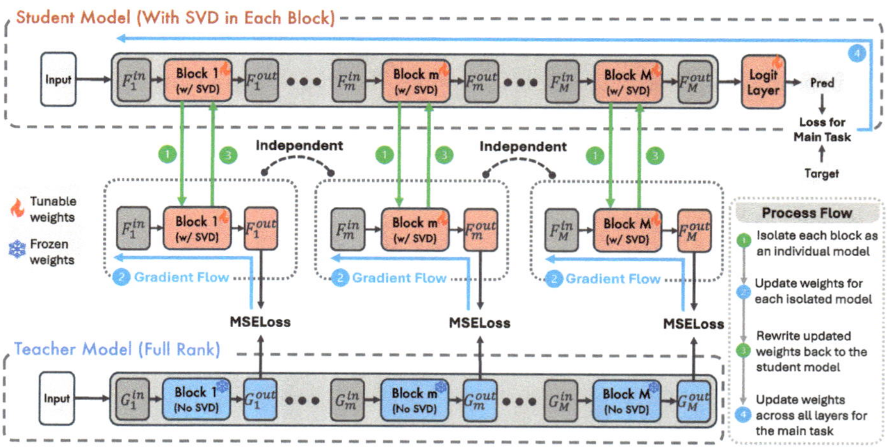

Fig. 7.2 Illustration of the layer-wise low-rank SVD KD process. All layers in the student model are isolated: (1) each block is treated as an individual model; (2) weights for each block are updated by comparing its output to the matching teacher block; (3) updated weights are written back to the student block; (4) a final global fine-tuning is performed

where U and V are orthogonal, and Σ is diagonal with singular values in descending order. Retaining only the top r singular values and their corresponding vectors produces a low-rank approximation, $\mathcal{W}_r = U_r \Sigma_r V_r^T$, which becomes the student model's weight.

While SVD captures most of the important structure with fewer parameters, SVD alone may not fully preserve the fine-grained information required for complex tasks, especially in deep networks. To address this, we apply block-by-block knowledge distillation (KD).

7.2.3 Block-by-Block Knowledge Distillation

Following SVD, we align each student block's output with its teacher counterpart using block-by-block KD. This involves isolating blocks by preventing gradient flow between them, so KD is applied independently to each:

- **Isolation:** Each block is treated as an independent model. For block m, input F_m^{in} yields output F_m^{out}, which is compared to G_m^{out} of the teacher.
- **Block loss:** The block KD loss is

$$\mathcal{L}_{\text{KD}}^{(m)} = \| F_m^{\text{out}} - G_m^{\text{out}} \|^2 \tag{7.2}$$

  and is minimized by updating only the weights within that block.

Block losses and updates are computed and applied independently for each block (see Fig. 7.3).

After all blocks are distilled individually, their weights are re-integrated into the student. A final global fine-tuning step with the main task objective (e.g., cross-entropy loss) is then performed to polish the full model. Because each block's outputs are closely aligned with the teacher, error accumulation is reduced and learning is stabilized, especially for deep architectures.

7.2.4 Comparison: Unified KD and Gradient Accumulation

Traditional knowledge distillation (unified KD) supervises all blocks using a global loss across the full model:

$$\mathcal{L}_{\text{unified-KD}} = \sum_{m=1}^{M} \mathcal{L}_{\text{KD}}^{(m)} \tag{7.3}$$

where each $\mathcal{L}_{\text{KD}}^{(m)}$ is as defined in (7.2). This results in cumulative gradients flowing from all later layers to earlier ones, as illustrated in Fig. 7.3.

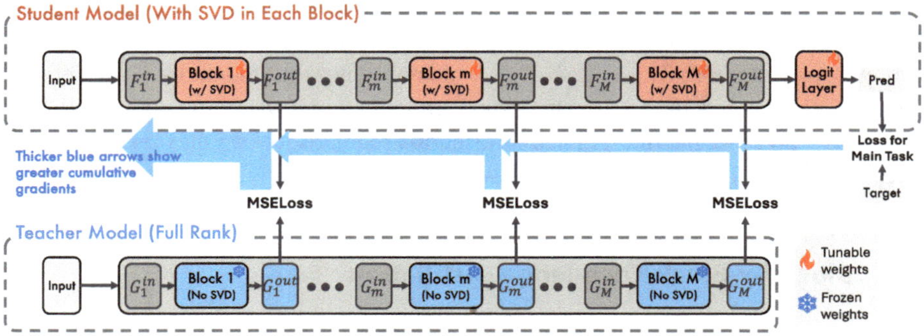

Fig. 7.3 Illustration of traditional unified KD with intermediate feature loss: earlier layers receive cumulative gradients from later layers (thicker blue arrows), which can destabilize learning and lead to inefficient alignment, especially in deep models

The gradient update for the weights of an earlier block m is

$$\frac{\partial \mathcal{L}_{\text{unified-KD}}}{\partial \mathcal{W}_m} = \sum_{k=m+1}^{M} \frac{\partial \mathcal{L}_{\text{KD}}^{(k)}}{\partial F_k^{\text{out}}} \frac{\partial F_k^{\text{out}}}{\partial F_{k-1}^{\text{out}}} \cdots \frac{\partial F_{m+1}^{\text{out}}}{\partial \mathcal{W}_m} \tag{7.4}$$

where the summation represents back-propagation of gradients from all higher blocks. As a result, earlier blocks experience gradient influences from later losses, leading to possible instability and less efficient learning—challenges our block-by-block method is designed to overcome.

7.3 Experiments

Dataset

We evaluate our method on a range of datasets for vision and language models, across various tasks. For vision models, we use Tiny-ImageNet [16], CIFAR-10 [14], CIFAR-100 [14], and SVHN [20]. Tiny-ImageNet consists of 64 × 64 images, while CIFAR-10, CIFAR-100, and SVHN have 32 × 32 images. For language models, we use the IMDB dataset [19], a standard benchmark for sentiment analysis.

7.3.1 Training Setup

All models were trained on two Nvidia GeForce RTX 4090 24GB GPUs.

ViT-Based Models (i.e., ViT∗ [4],[1] Swin Transformer [18], and CaiT [25]) used AdamW with a batch size of 256, learning rate of 0.001, and weight decay of 5e-2. In block-by-block KD, we adjust the learning rate to 0.01.

CNN Model ResNet [6][2] was trained with stochastic gradient descent (SGD), a batch size of 256, learning rate of 0.1, weight decay of 1e-4, and momentum of 0.9. There are no changes in block-by-block KD.

LLMs[3] GPT-2 [23] was trained using AdamW, batch size 8, and learning rate 5e-6. DeBERTa [8] was trained with a batch size of 16 and learning rate of 5e-6. In block-by-block KD, we increase the learning rate for both models to 5e-5. Cross-entropy loss was used as the main task loss for all models.

Teacher and Student Models in Block-by-Block KD

In our setup, the teacher model is a full-rank model without SVD compression. The student model is formed by applying SVD to the fully-connected and convolutional layers, except for the final logit layer, of the teacher. For block-by-block KD, each block (Transformer or Bottleneck) has a separate, independent optimizer instance that is of the same type as the entire student model (e.g., AdamW for ViT models and LLMs, SGD for ResNet). All hyperparameters match those of the student model. Learning rater were increased by 10× (except for ResNet, where the learning rate was unchanged) to accelerate convergence and help each student block better approximate its teacher. We use MSE loss to measure the difference between the outputs of each isolated student block and its corresponding teacher block.

7.3.2 Block-by-Block KD Boosts Model Performance

We evaluate the effectiveness of block-by-block KD in improving performance for different model architectures across a variety of tasks.

[1] ViT∗ specifically refers to architecture from [4], and **not** the original ViT architecture from [3]. We use the settings from the official implementation of [4]: https://github.com/hananshafi/vits-for-small-scale-datasets/tree/main.

[2] We use the settings from the official implementation: https://github.com/pytorch/examples/tree/main/imagenet.

[3] We use settings from the Hugging Face implementation: https://github.com/huggingface/transformers/tree/main/docs/source/en/model_doc.

Table 7.1 Top-1 accuracy (%) of CNN models on Tiny-ImageNet datasets. The symbol † denotes re-implemented results. Our block-by-block KD method improves the alignment between student and teacher weights, resulting in performance boosts. The rank of each layer of full rank ResNeT varies based on the number of channels

Model	#Param (tunable)	Rank	Tiny-ImageNet
ResNet-18†	11.28M (11.28M)	Full	56.62
ResNet-50†	23.92M (23.92M)	Full	63.63
ResNet-50	10.32M (10.05M)	128	61.61
ResNet-50 + block-by-block KD	10.32M (10.05M)	128	**62.74**
ResNet-50	5.50M (5.23M)	64	38.66
ResNet-50 + block-by-block KD	5.50M (5.23M)	64	**59.56**
ResNet-50	3.10M (2.82M)	32	5.73
ResNet-50 + block-by-block KD	3.10M (2.82M)	32	**48.33**

The bold values highlight the best configuration for a given rank, model and test set configuration.

Compressed CNN Model

Table 7.1 demonstrates the effectiveness of block-by-block KD in enhancing compressed CNNs, specifically ResNet-50. At rank 128, ResNet-50 has only 10.32M parameters, smaller than ResNet-18's full-rank model (11.28M), yet achieves a higher accuracy of 61.61% on Tiny-ImageNet. With block-by-block KD, ResNet-50's accuracy further improves to 62.74%, approaching the full-rank teacher model's performance (63.63%).

At a much lower rank 32, ResNet-50's accuracy drops significantly to 5.73% with only SVD-based compression. With block-by-block KD, it can still maintain 48.33% accuracy. This provides evidence that block-by-block KD can improve robustness in low-rank settings, as stronger alignment between student and teacher weights may compensate for aggressive parameter reduction.

Compressed LLM Models

Table 7.2 shows the effectiveness of block-by-block KD in improving compressed LLM models (i.e., GPT-2 and DeBERTa) on the IMDB dataset. Across both models, block-by-block KD consistently enhances performance. For instance, at a rank of 128, model sizes are significantly reduced to 19M for GPT-2 (4.5× smaller) and 22M for DeBERTa (4× smaller), excluding the fixed word embedding layers (39.4M for GPT-2 and 98.4M for DeBERTa), which are not part of the SVD compression. For simplicity, only the compressed model sizes without word embedding layers are reported. Without block-by-block KD, performance drops to 83.75% and 83.56% for GPT-2 and DeBERTa, respectively. With block-by-block KD, the accuracy improves to 88.56% for GPT-2 and 90.99% for DeBERTa, closely matching the performance of the full-rank models.

With more aggressive compression at rank 16, model size further drops to 2.5M for GPT-2 (34× smaller) and 3.2M for DeBERTa (27× smaller). Without block-by-block KD, SVD-only compressed GPT-2 and DeBERTa achieve accuracies of 82.48% and

Table 7.2 Top-1 accuracy (%) of LLMs on IMDB datasets. Model sizes exclude word embeddings. Our block-by-block KD method improves the alignment between student and teacher weights, resulting in performance boosts

Model	#Param (tunable)	Rank	IMDB
GPT-2	85.05M (85.05M)	Full (768)	88.88
GPT-2	19.00M (18.96M)	128	83.75
GPT-2 + block-by-block KD	19.00M (18.96M)	128	**88.56**
GPT-2	9.56M (9.52M)	64	83.54
GPT-2 + block-by-block KD	9.56M (9.52M)	64	**87.74**
GPT-2	4.84M (4.80M)	32	83.64
GPT-2 + block-by-block KD	4.84M (4.80M)	32	**86.44**
GPT-2	2.48M (2.44M)	16	82.48
GPT-2 + block-by-block KD	2.48M (2.44M)	16	**83.57**
DeBERTa	86.04M (86.04M)	Full (768)	92.01
DeBERTa	21.96M (21.53M)	128	83.56
DeBERTa + block-by-block KD	21.96M (21.53M)	128	**90.99**
DeBERTa	11.24M (10.80M)	64	83.33
DeBERTa + block-by-block KD	11.24M (10.80M)	64	**90.36**
DeBERTa	5.88M (5.44M)	32	82.80
DeBERTa + block-by-block KD	5.88M (5.44M)	32	**89.40**
DeBERTa	3.20M (2.77M)	16	81.79
DeBERTa + block-by-block KD	3.20M (2.77M)	16	**87.58**

The bold values highlight the best configuration for a given rank, model and test set configuration.

81.79%, respectively. With block-by-block KD, performance rises to 83.57% for GPT-2 and 87.58% for DeBERTa.

Compressed ViT-Based Models

Table 7.3 illustrates performance improvements achieved by block-by-block KD for compressed ViT-based models (ViT, Swin Transformer, and CaiT) across multiple datasets. For ViT, the full-rank version with 2.77M parameters achieves an accuracy of 64.06% on Tiny-ImageNet and 96.80% on SVHN. Compressed to rank 64, ViT's parameters reduce to 1.45M (1.38M tunable), with accuracies dropping to 59.77% on Tiny-ImageNet and 96.36% on SVHN. With block-by-block KD, the rank-64 ViT model improves to 60.94% on Tiny-ImageNet and 96.74% on SVHN. At an even higher compression level with rank 16, the ViT model reduces further to 0.45M parameters (0.38M tunable), though accuracy declines significantly to 35.55% on Tiny-ImageNet and 38.84% on SVHN. By applying block-by-block KD, the rank-16 ViT model achieves 44.14% on Tiny-ImageNet and 92.80% on SVHN.

Table 7.3 Top-1 accuracy (%) of ViT-based models on Tiny-ImageNet, CIFAR-10, CIFAR-100, and SVHN datasets. The symbol † denotes re-implemented results. If there are differences between results within a dataset and given rank, the best results are **bolded**. Block-by-block KD better aligns and teacher weights, resulting in improved performance

Model	#Param (tunable)	Rank	Tiny-ImageNet	CIFAR-10	CIFAR-100	SVHN
ViT*†	2.77M (2.77M)	Full	64.06	96.88	80.86	96.80
ViT*	1.45M (1.38M)	64	59.77	95.70	73.40	96.36
ViT* + block-by-block KD	1.45M (1.38M)	64	**60.94**	**96.09**	**74.61**	**96.74**
ViT*	0.78M (0.72M)	32	42.97	77.73	51.17	91.54
ViT* + block-by-block KD	0.78M (0.72M)	32	**51.56**	**88.28**	**62.11**	**95.97**
ViT*	0.45M (0.38M)	16	35.55	42.27	30.08	38.84
ViT* + block-by-block KD	0.45M (0.38M)	16	**44.14**	**78.91**	**46.88**	**92.80**
Swin†	7.13M (7.13M)	Full	70.31	95.31	80.86	97.86
Swin	2.45M (2.42M)	64	60.94	93.36	75.00	97.49
Swin + block-by-block KD	2.45M (2.42M)	64	**67.58**	**94.14**	**76.56**	**97.59**
Swin	1.28M (1.26M)	32	43.75	81.25	51.17	92.45
Swin + block-by-block KD	1.28M (1.26M)	32	**54.30**	**88.67**	**67.19**	**96.18**
Swin	0.70M (0.68M)	16	23.44	46.09	21.88	44.29
Swin + block-by-block KD	0.70M (0.68M)	16	**36.33**	**73.83**	**44.53**	**91.03**
CaiT†	8.12M (8.12M)	Full	66.80	94.92	81.25	97.93
CaiT	4.14M (4.06M)	64	62.50	94.53	69.53	**97.86**
CaiT + block-by-block KD	4.14M (4.06M)	64	**67.97**	94.53	**74.61**	97.78
CaiT	2.15M (2.07M)	32	39.06	91.80	58.98	96.78
CaiT + block-by-block KD	2.15M (2.07M)	32	**63.28**	91.80	**67.97**	**97.32**
CaiT	1.15M (1.07M)	16	21.09	68.75	29.69	86.29
CaiT + block-by-block KD	1.15M (1.07M)	16	**48.44**	**84.77**	**53.52**	**95.95**

Similarly, the Swin Transformer exhibits notable gains with block-by-block KD. The full-rank model has 7.13M parameters and achieves 70.31% accuracy on Tiny-ImageNet and 97.86% on SVHN. Compressed to rank 64, Swin Transformer's size is reduced to 2.45M (2.42M tunable), yielding 60.94% on Tiny-ImageNet and 97.49% on SVHN. With block-by-block KD, the rank-64 model improves significantly, reaching 67.58% on Tiny-ImageNet and 97.59% on SVHN. At the lowest compression setting, rank 16, Swin Transformer compresses to 0.70M parameters (0.68M tunable), with accuracies of 23.44% on Tiny-ImageNet and 44.29% on SVHN. Block-by-block KD substantially boosts these values to 36.33% on Tiny-ImageNet and 91.03% on SVHN, showing the effectiveness of block-by-block KD in maintaining performance under extreme low-rank compression.

Table 7.4 Ablation study comparing traditional unified KD with our block-by-block KD approach. Model sizes exclude word embeddings

Model	KD	#Param (Tunable)	Rank	IMDB
DeBERTa	None	86.04M (86.04M)	Full (768)	92.01
DeBERTa	None	21.96M (21.53M)	128	83.56
DeBERTa	Unified	21.96M (21.53M)	128	88.08 (+4.52)
DeBERTa	Block-by-block	21.96M (21.53M)	128	**90.99** (+7.43)
DeBERTa	None	11.24M (10.80M)	64	83.33
DeBERTa	Unified	11.24M (10.80M)	64	87.25 (+3.92)
DeBERTa	Block-by-block	11.24M (10.80M)	64	**90.36** (+7.03)
DeBERTa	None	5.88M (5.44M)	32	82.80
DeBERTa	Unified	5.88M (5.44M)	32	86.47 (+3.67)
DeBERTa	Block-by-block	5.88M (5.44M)	32	**89.40** (+6.6)
DeBERTa	None	3.20M (2.77M)	16	81.79
DeBERTa	Unified	3.20M (2.77M)	16	84.88 (+3.09)
DeBERTa	Block-by-block	3.20M (2.77M)	16	**87.58** (+5.79)

The bold values highlight the best configuration for a given rank, model and test set configuration.

7.3.3 Ablation Study

Comparison to Unified KD

Table 7.4 evaluates the effectiveness of the proposed approach compared to traditional unified KD, we ablate DeBERTa models compressed to various ranks. Word embedding layers are excluded from model size calculations as they remain uncompressed by SVD. For unified KD, the total loss is defined as

$$\mathcal{L}_{\text{total}} = \mathcal{L}_{\text{cross-entropy}} + \lambda \sum_{m=1}^{M} \mathcal{L}_{\text{KD}}^{(m)}, \tag{7.5}$$

where $\mathcal{L}_{\text{KD}}$ is the MSE-based distillation loss applied to the outputs of each block m across M blocks, as defined in Eq. (7.3). We set λ to be 10. The full-rank DeBERTa model (86.04M parameters) achieves the highest accuracy on the IMDB dataset, reaching 92.01%. At rank 128, DeBERTa without KD (SVD only) shows a decrease in performance to 83.56%. Unified KD improves accuracy to 88.08%, while block-by-block KD further boosts it to 90.99%, achieving a 7.43% boost over the SVD-only model and closely matching the full-rank performance. At rank 16, DeBERTa with SVD only shows the lowest accuracy, 81.79%, with a compressed model size of 3.20M. Unified KD improves this to 84.88%, and block-by-block KD further enhances accuracy to 87.58%, achieving a 5.79% boost over the SVD-only model. These results demonstrate the superior effectiveness of block-by-block KD across different compression levels.

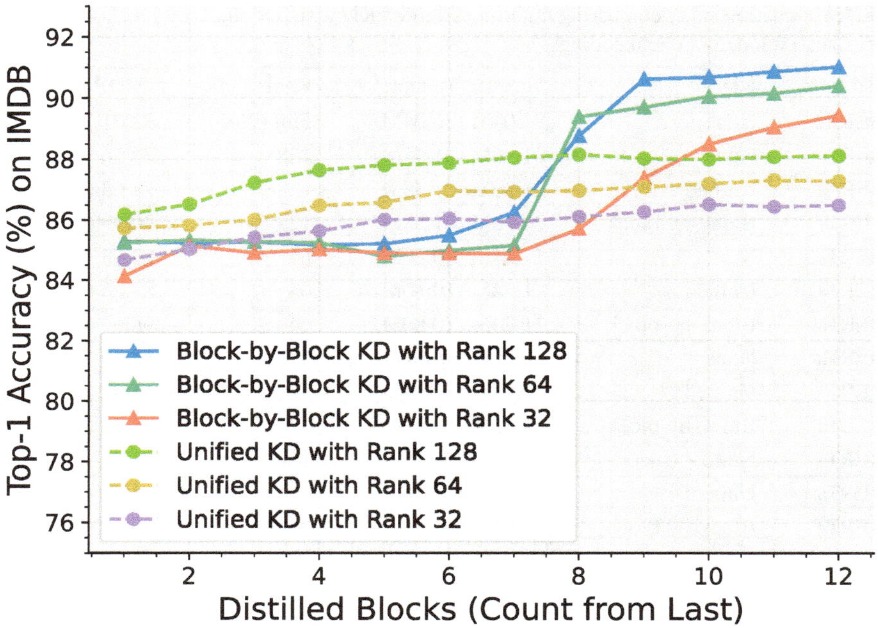

Fig. 7.4 Top-1 accuracy on IMDB for DeBERTa (12 Transformer blocks) compressed to ranks 128, 64, and 32, as the number of distilled blocks increases from the last block. Block-by-block KD (triangles) shows significant performance gains, especially as more blocks are included. Unified KD (circles) shows only minor gains when distilling into earlier blocks, highlighting limitations due to the compounded gradient effect

Effect of Distilling Blocks

Figure 7.4 illustrates the impact of distilling progressively more blocks in DeBERTa, which has 12 Transformer blocks, on the IMDB dataset, compressed to low ranks of 128, 64, and 32. The comparison highlights the differences between block-by-block KD and unified KD. In unified KD, distilling later blocks (from the last up to the 6th) steadily improves performance. Gains become marginal as earlier blocks are added, which indicates that including more blocks, especially the earliest ones, yields limited benefit.

This trend reflects the compounded gradient effect, where cumulative gradients from later blocks repeatedly affect earlier blocks, as shown by Eq. (7.4), resulting in inefficient learning. As more blocks are distilled, earlier blocks receive multiple, sometimes conflicting, gradient updates, which can lead to instability.

In contrast, block-by-block KD consistently shows significant performance gains, even when performing distillation with the earliest blocks of a model. This suggests that allowing each block to learn independently and avoid inter-block gradient interference aligns student and teacher features more effectively.

Feature Alignment Visualization

To assess how well block-by-block KD aligns student features with those of the teacher model, we use Smooth Grad-CAM++ [21] to visualize the focus regions for both models. Figure 7.5 shows activation maps for ResNet-50 at rank 64, focusing on *layer2*. Each set of images includes, from left to right, the original input, followed by focus regions from the full-rank teacher model, block-by-block KD (Ours), and the SVD-only models. We observe that block-by-block KD helps the student model focus on regions similar to those highlighted by the teacher, indicating strong feature alignment. Meanwhile, the SVD-only model focuses on less relevant regions, showing a lack of alignment without the targeted adjustments provided by block-by-block KD.

7.4 Discussion and Future Work

The block-by-block knowledge distillation (KD) framework introduced in this chapter offers a modular and parallelizable approach to model compression. By isolating each structural block as an independent unit for distillation, the method not only improves student-teacher alignment but also lays the groundwork for efficient distributed and federated learning. Because each block can be updated in isolation, block-wise distillation is well-suited for parallel updates on memory-constrained or heterogeneous devices, where only a subset of the model may be loaded or updated at any one time.

Nevertheless, deploying such a block-wise approach in real-world distributed scenarios introduces new challenges. Communication overhead may increase due to frequent weight transfers between devices, and ensuring model consistency under asynchronous updates remains an open question. Furthermore, device hardware and memory constraints may necessitate adaptation of block granularity or size, motivating further research into adaptive block partitioning and distributed KD protocols.

Another promising avenue for future work is the adoption of variable-rank compression across model layers. While our experiments used a constant rank across all blocks for simplicity, an analysis of the singular value spectra (Fig. 7.6) in DeBERTa reveals considerable variation in the "elbow point"—the index at which singular values decline sharply—across layers. Most layers display an elbow index well below 20, while some, such as dense output layers, exhibit much higher elbow indices (up to 766). This suggests that layer-wise or block-wise selection of rank, guided by the singular value distribution, may yield improved compression-accuracy trade-offs. Systematically exploring variable rank selection represents a valuable direction for future study. Adaptive rank selection could yield even better compression–accuracy trade-offs.

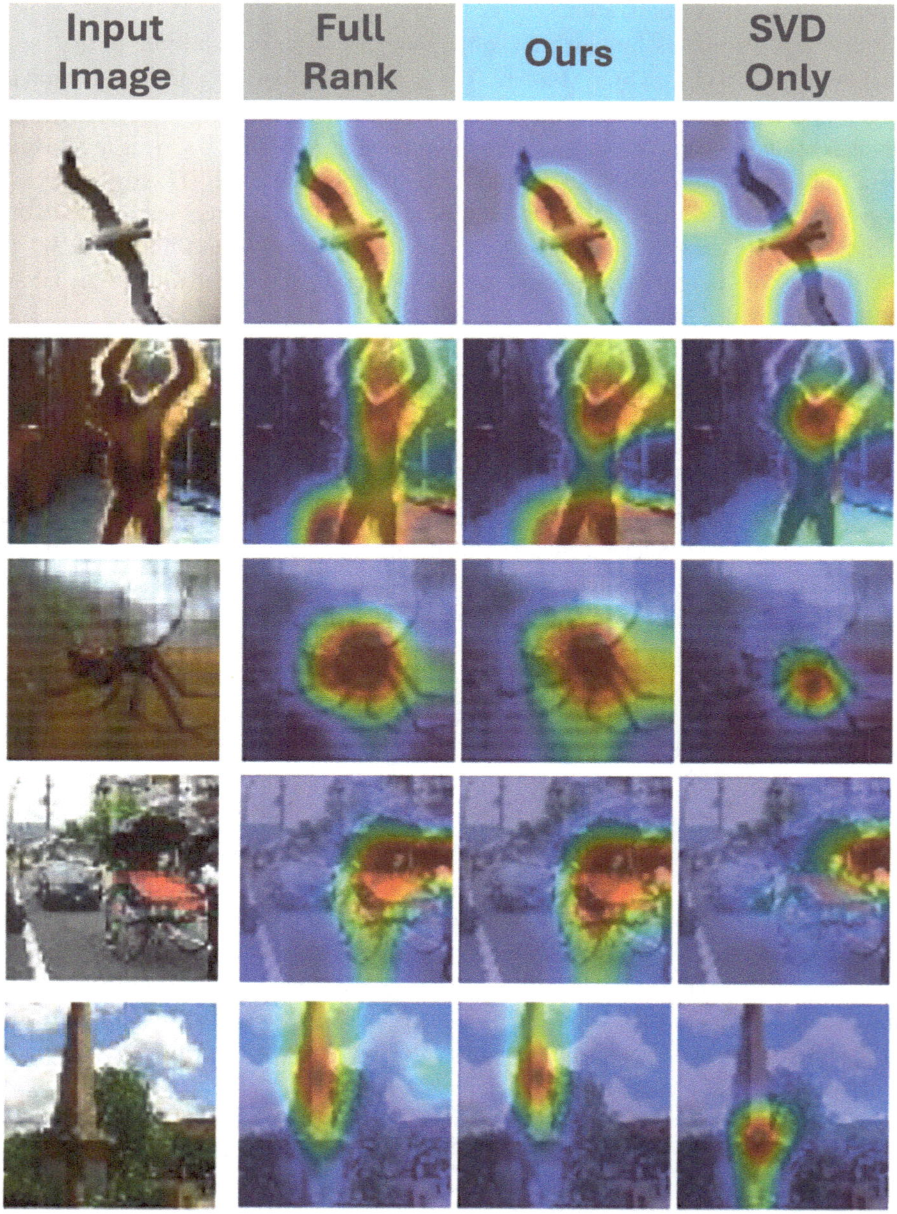

Fig. 7.5 Smooth Grad-CAM++ visualization for ResNet-50 at rank 64 on *layer2*. From left: input, full-rank teacher, block-by-block KD (Ours), and SVD-only models. Block-by-block KD aligns student focus with teacher regions

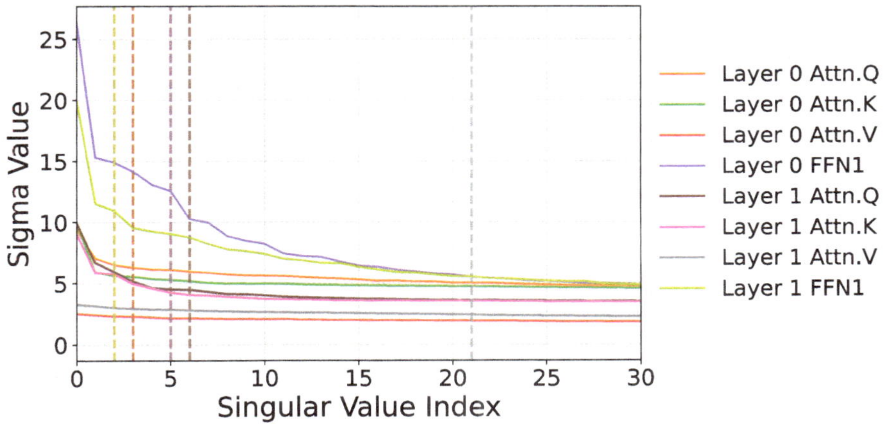

Fig. 7.6 Each curve represents the singular values from the matrices of the first two layers of DeBERTa. The dashed vertical lines indicate the elbow points for each matrix. Most elbow indices are below 20, with the notable exception of the dense output layers, which have an elbow index of 766

7.5 Summary

The proposed block-by-block knowledge distillation (KD) framework provides a robust and versatile strategy for compressing deep neural networks via localized, low-rank approximations. By focusing distillation at the level of network blocks, this method overcomes the compounded optimization and gradient propagation difficulties that often arise in unified, end-to-end distillation schemes. This granular approach enables more stable training and more faithful transfer of internal representations from a high-capacity teacher to a compact student model.

Extensive experiments demonstrate both the generality and effectiveness of this approach across diverse neural architectures including convolutional networks, vision transformers, and language models. Block-by-block KD enables aggressive rank reduction and model compression. Remarkably, reducing the SVD rank from 768 to 16 yields up to a 27× decrease in model size, often with less than a 5% loss in accuracy, showing that the framework can retain most of the teacher's performance even under severe constraints.

Together, these results establish block-by-block KD as a powerful model compression paradigm, with broad applicability in scenarios requiring efficient neural network deployment. Future efforts may build on this foundation to further enhance resource efficiency, scalability, and adaptability in practical deep learning systems. Given the results in the plots of the singular values, it is likely possible to extract further compression through rank truncation.

References

1. Tom B. Brown, Benjamin Mann, Nick Ryder, Melanie Subbiah, Jared Kaplan, Prafulla Dhariwal, Arvind Neelakantan, Pranav Shyam, Girish Sastry, Amanda Askell, Sandhini Agarwal, Ariel Herbert-Voss, Gretchen Krueger, Tom Henighan, Rewon Child, Aditya Ramesh, Daniel M. Ziegler, Jeffrey Wu, Clemens Winter, Christopher Hesse, Mark Chen, Eric Sigler, Mateusz Litwin, Scott Gray, Benjamin Chess, Jack Clark, Christopher Berner, Sam McCandlish, Alec Radford, Ilya Sutskever, and Dario Amodei. Language models are few-shot learners, 2020.
2. Yu Cheng, Duo Wang, Pan Zhou, and Tao Zhang. A survey of model compression and acceleration for deep neural networks, 2020.
3. Alexey Dosovitskiy, Lucas Beyer, Alexander Kolesnikov, Dirk Weissenborn, Xiaohua Zhai, Thomas Unterthiner, Mostafa Dehghani, Matthias Minderer, Georg Heigold, Sylvain Gelly, Jakob Uszkoreit, and Neil Houlsby. An image is worth 16x16 words: Transformers for image recognition at scale, 2021.
4. Hanan Gani, Muzammal Naseer, and Mohammad Yaqub. How to train vision transformer on small-scale datasets? In *33rd British Machine Vision Conference 2022, BMVC 2022, London, UK, November 21–24, 2022*. BMVA Press, 2022.
5. Song Han, Huizi Mao, and William J. Dally. Deep compression: Compressing deep neural networks with pruning, trained quantization and Huffman coding, 2016.
6. Kaiming He, Xiangyu Zhang, Shaoqing Ren, and Jian Sun. Deep residual learning for image recognition, 2015.
7. Kaiming He, Xiangyu Zhang, Shaoqing Ren, and Jian Sun. Identity mappings in deep residual networks, 2016.
8. Pengcheng He, Jianfeng Gao, and Weizhu Chen. DeBERTaV3: Improving deberta using electra-style pre-training with gradient-disentangled embedding sharing, 2023.
9. Ruifei He, Shuyang Sun, Jihan Yang, Song Bai, and Xiaojuan Qi. Knowledge distillation as efficient pre-training: Faster convergence, higher data-efficiency, and better transferability, 2022.
10. Byeongho Heo, Jeesoo Kim, Sangdoo Yun, Hyojin Park, Nojun Kwak, and Jin Young Choi. A comprehensive overhaul of feature distillation. In *ICCV*, pages 1921–1930, 2019.
11. Byeongho Heo, Minsik Lee, Sangdoo Yun, and Jin Young Choi. Knowledge transfer via distillation of activation boundaries formed by hidden neurons. In *AAAI*, pages 3779–3787, 2019.
12. Geoffrey Hinton, Oriol Vinyals, and Jeff Dean. Distilling the knowledge in a neural network, 2015.
13. Jangho Kim, Seonguk Park, and Nojun Kwak. Paraphrasing complex network: Network compression via factor transfer. In *NeurIPS*, pages 2765–2774, 2018.
14. Alex Krizhevsky. Learning multiple layers of features from tiny images. Technical report, University of Toronto, 2009.
15. Alex Krizhevsky, Ilya Sutskever, and Geoffrey E Hinton. Imagenet classification with deep convolutional neural networks. In F. Pereira, C.J. Burges, L. Bottou, and K.Q. Weinberger, editors, *Advances in Neural Information Processing Systems*, volume 25. Curran Associates, Inc., 2012.
16. Ya Le and Xuan Yang. Tiny imagenet visual recognition challenge. *CS 231N*, 7(7):3, 2015.
17. Yann LeCun, Yoshua Bengio, and Geoffrey Hinton. Deep learning. *Nature*, 521(7553):436, 2015.
18. Ze Liu, Yutong Lin, Yue Cao, Han Hu, Yixuan Wei, Zheng Zhang, Stephen Lin, and Baining Guo. Swin transformer: Hierarchical vision transformer using shifted windows, 2021.

19. Andrew L. Maas, Raymond E. Daly, Peter T. Pham, Dan Huang, Andrew Y. Ng, and Christopher Potts. Learning word vectors for sentiment analysis. In Dekang Lin, Yuji Matsumoto, and Rada Mihalcea, editors, *Proceedings of the 49th Annual Meeting of the Association for Computational Linguistics: Human Language Technologies*, pages 142–150, Portland, Oregon, USA, June 2011. Association for Computational Linguistics.
20. Yuval Netzer, Tao Wang, Adam Coates, Alessandro Bissacco, Bo Wu, and Andrew Y. Ng. Reading digits in natural images with unsupervised feature learning. In *NIPS Workshop on Deep Learning and Unsupervised Feature Learning 2011*, 2011.
21. Daniel Omeiza, Skyler Speakman, Celia Cintas, and Komminist Weldemariam. Smooth Grad-CAM++: An enhanced inference level visualization technique for deep convolutional neural network models. *CoRR*, abs/1908.01224, 2019.
22. Alec Radford and Karthik Narasimhan. Improving language understanding by generative pre-training. OpenAI Technical Report, 2018. Available at https://cdn.openai.com/research-covers/language-unsupervised/language_understanding_paper.pdf.
23. Alec Radford, Jeff Wu, Rewon Child, David Luan, Dario Amodei, and Ilya Sutskever. Language models are unsupervised multitask learners. OpenAI Technical Report, 2019. Available at https://cdn.openai.com/better-language-models/language_models_are_unsupervised_multitask_learners.pdf.
24. Adriana Romero, Nicolas Ballas, Samira Ebrahimi Kahou, Antoine Chassang, Carlo Gatta, and Yoshua Bengio. Fitnets: Hints for thin deep nets, 2015.
25. Hugo Touvron, Matthieu Cord, Alexandre Sablayrolles, Gabriel Synnaeve, and Hervé Jégou. Going deeper with image transformers, 2021.
26. Tianhe Yu, Saurabh Kumar, Abhishek Gupta, Sergey Levine, Karol Hausman, and Chelsea Finn. Gradient surgery for multi-task learning, 2020.

8 Leveraging Cloud Knowledge Locally: Lightweight Edge-Side Classification with Cloud-Extracted Features

Chapter Overview

The preceding chapters have focused on optimizing the "Matrix-Centric" stack within a single execution environment. However, real-world deployment often involves a distributed hierarchy: a fast but resource-limited **edge device** and a powerful but remote **cloud server**.

The central challenge addressed in this chapter is how to distribute the computational workload between these two entities. As neural networks continue to grow in size and complexity, it becomes increasingly challenging for small, resource-constrained devices to keep pace, especially when user privacy and personalization are paramount. This chapter introduces a new approach to model splitting that allows edge devices to benefit from the representational power of large cloud-based models, while keeping user data local and private.

Unlike traditional edge-cloud model partitioning—where feature extraction also takes place on the edge, our method shifts the heavy lifting to the cloud. The cloud is responsible for extracting high-quality features using its powerful models, while edge devices focus solely on maintaining and fine-tuning compact, task-specific classifiers. This not only reduces the compute and memory burden on the user device, but also simplifies and accelerates personalization with local data as the cloud does not need to track user personalization.

Crucially, our framework is designed with privacy in mind: user data labels never leave the edge, and the cloud provider's proprietary models remain protected. To further support generalization and rapid adaptation, the cloud can provide "anchors", prototypical feature representations to bootstrap or enhance local classifiers, but the user retains full control.

Through experiments on datasets such as CIFAR-100 and Food-101, we demonstrate that this division of labor enables accurate, efficient, and low-cost learning on the

A. Sabot, *Minimizing Data Movement and Parameter Count Across the Machine Learning Stack*, Synthesis Lectures on Computer Science,
https://doi.org/10.1007/978-3-032-23100-0_8

edge without sacrificing performance. Moreover, users can easily train or swap between multiple classifiers to address a variety of tasks, all while controlling their data and computational footprint.

This chapter offers an approach for implementing personalized and privacy-respecting AI on everyday devices.

8.1 Introduction

Modern vision models—particularly Vision Transformers (ViT) [5]—achieve state-of-the-art accuracy, yet their hundreds of millions of parameters and gigaflop-level inference cost place them far beyond the reach of mobile phones, AR/VR headsets, and smart sensors. At the same time, these edge devices are precisely where many emerging applications reside: retail-shelf monitoring, industrial defect detection, home robotics, and countless personal or context-aware services. The tension between the richness of cloud-scale models and the limitations of edge hardware motivates a hybrid paradigm in which the cloud supplies high-quality feature representations, while the edge performs only the final, task-specific classification.

Why Not Parameter-Efficient Fine-Tuning (PEFT)? Techniques such as LoRA and adapters [6, 18] reduce the number of *trainable* parameters, yet they still require the *entire* backbone to reside on-device during inference. For many edge platforms this remains prohibitive in latency, energy, and memory.

Our Premise Keep the large backbone in the cloud, transmit only its compact embedding vectors ($\approx 10^2 - 10^3$ floats), and let the edge learn a tiny classifier. Because the final layers are delegated to the client, each device can personalize its model with minimal compute, storage, or privacy cost: labels stay local, and only a few kilobytes of weights are updated.

Early-Exit vs. Late-Start Prior "early-exit" networks (e.g. BranchyNet [22]) attempt to stop computation early on the edge. We take the opposite stance: all heavy feature extraction occurs in the cloud, and the edge performs the last $\sim$1–2 layers—a *late-start* rather than an early-exit. This inversion enables:

- **Tiny on-device models**: a linear or shallow MLP head.
- **Rapid personalisation**: training completes in seconds.
- **Continual learning**: new classes or conditions can be incorporated without re-running or re-deploying the cloud backbone.
- **Privacy**: only abstract features—not raw images or sensor data—leave the device.

Key Observation—Centroids Suffice Feature vectors produced by large ViT backbones exhibit strong class structure [7]. We show that, once features are extracted, a *single* class centroid per class can serve as a "knowledge anchor" that nearly matches the accuracy obtained from the full training set. Sending only these centroids from cloud to edge reduces both communication and local storage.

Chapter Roadmap Section 8.2 formalizes the split-model system and illustrates practical use cases. Section 8.3 revisits supervised learning with frozen feature extractors. Section 8.6 differentiates between raw datasets and *featuresets*, and explains how centroids are created. Section 8.8 details the local training procedure, and Sect. 8.9 empirically evaluates accuracy, latency, and memory on vision benchmarks. We discuss limitations and deployment considerations in Sect. 8.10, and position our work within the broader literature in Sect. 8.4.

Contributions

1. We propose a cloud/edge split in which only a lightweight classifier runs locally while the cloud provides frozen embeddings.
2. We demonstrate that class centroids—one or a few per class—are sufficient to train or re-train the local head, enabling cheap continual learning.
3. We show how centroids from multiple datasets can be merged to distil knowledge across tasks without accessing original data.

8.2 Overview and Example Scenarios

This chapter studies a *split-model* training pipeline in which a high-capacity *cloud* feature extractor works in tandem with a small, adaptive *client* classifier (Fig. 8.1). The setting we target is common in modern IoT and mobile applications:

- Client devices have a reliable network connection but are heavily constrained in memory, compute, and energy (e.g. phones, smart cameras, AR/VR headsets).
- The cloud can host large Vision-Transformer (ViT) backbones, but re-training such models for every user or task would be prohibitively slow and costly.

High-Level Workflow

1. The cloud model F receives raw images from the client and returns fixed feature vectors ("embeddings").
2. The client trains, updates, or extends a lightweight classifier C on these embeddings using its own labels.

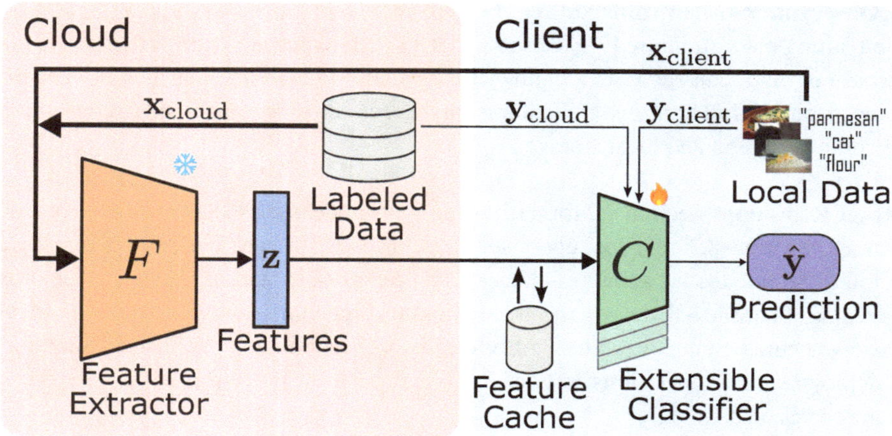

Fig. 8.1 Proposed "cloud-to-edge" workflow. A fixed, high-capacity cloud model extracts embedding vectors from user data. These embeddings—which are far smaller and more abstract than raw inputs—are cached on the device. A lightweight on-device classifier is then trained (or re-trained) locally using a small, labelled *featureset*. The cloud can optionally send additional synthetic anchors (class centroids) to stabilize accuracy when multiple datasets or new classes are added

3. For future inputs, the client can either (i) send the image for cloud embedding and run C locally, or (ii) employ a local emulator of F [23] if full privacy is required.

Rather than store full image datasets, the client caches only a handful of embeddings—sometimes as few as a single *class centroid* per class. Our experiments (Sect. 8.9) show that these tiny *featuresets* are sufficient to re-train the classifier in seconds, making on-device personalisation practical.

8.2.1 Example Scenario 1: Defect Detection

In semiconductor fabrication, dozens of high-resolution cameras inspect wafers at multiple process stages. When a new defect type is discovered by an automated algorithm [24], deploying a completely new model to every camera is expensive. With the proposed split pipeline the cloud extracts embeddings once, then each camera updates *only* its final classifier row for the new defect. Retraining is therefore inexpensive and stage-specific.

8.2.2 Example Scenario 2: Smart Security Cameras

A home security camera may temporarily prioritize a specific person (e.g. a guest) or object (e.g. a delivery box). Adding such a transient class requires only appending a single row to the linear head and supplying one centroid embedding. When the object is no longer relevant, the row is removed and the classifier instantly reverts to its previous state—no cloud retraining needed.

8.3 Background

8.3.1 Large Transformers as Feature Extractors

Networks with wide layers learn remarkably similar early-layer representations, even when trained on different datasets [8]. Vision Transformers (ViTs) and related models generalize well to out-of-distribution data [25], motivating their use as universal feature extractors in our pipeline.

8.3.2 Supervised Learning Formalism

Let the labelled image dataset be $D = \{(\mathbf{x}_i, \mathbf{y}_i)\}_{i=1}^{N}$, with M classes and class-specific subsets $D_m = \{(\mathbf{x}_j^m, m)\}_{j=1}^{N_m}$. A feature extractor $F : \mathbf{x} \mapsto \mathbf{z} \in \mathbb{R}^f$ feeds a classifier $C : \mathbf{z} \mapsto \hat{\mathbf{y}}$. Training minimizes

$$\min_{\theta_F, \theta_C} \sum_{i=1}^{N} L\big(C(F(\mathbf{x}_i; \theta_F); \theta_C), \mathbf{y}_i\big),$$

using, e.g., stochastic gradient descent.

8.3.3 Frozen Feature Extractor

When F is fixed, each image maps deterministically to an embedding, yielding a *featureset* $Z = \{(\mathbf{z}_i, \mathbf{y}_i)\}_{i=1}^{N}$. Optimisation now concerns only θ_C, greatly reducing memory and compute. Linear probes [1] are a canonical instance of this paradigm.

8.3.4 Virtual Samples

A *virtual sample* $(\bar{\mathbf{z}}, \mathbf{y})$ is an embedding that did *not* originate from a real image but is synthetically assigned a label (e.g. a class centroid). Such virtual data can dramatically shrink the featureset size while maintaining accuracy, as we demonstrate in Sect. 8.9.

8.4 Related Work

8.4.1 Transfer Learning and Foundation Models

Transfer learning transfers knowledge from a source model or task to a target model or task [17, 21]. Recent "foundation models"—large networks pre-trained on web-scale data [26]—have become the de facto source models for a wide range of downstream applications. Parameter-efficient fine-tuning (PEFT) techniques such as adapters [18] and LoRA [6] update only a small fraction of parameters, yet they still require the entire backbone to reside on-device during inference. Our split pipeline instead keeps the backbone in the cloud and trains only a tiny head locally, eliminating the memory and compute cost of storing the full model at the edge.

8.4.2 Linear Probing and Frozen Feature Extractors

Linear probing trains a shallow classifier on frozen features to assess representation quality [1]. When a frozen backbone is good enough, linear probes can outperform full fine-tuning by avoiding feature drift [10]. Our work generalizes this idea: centroids and subcluster embeddings act as *virtual* samples, making training even lighter than a standard linear probe that would require all embeddings.

8.4.3 Federated Learning and Model Alignment

Federated learning (FL) trains a global model from distributed data [14,20]. SphereFed [4] and related work address alignment issues that arise when each client has its own feature extractor. In contrast, we *share* a single frozen extractor and allow each device to keep a private classifier. Although we do not exchange updates, our method could be combined with FL by aggregating the tiny classifier weights instead of full models.

8.4.4 Model Splitting and Early Exit

Model-splitting off-loads part of a network to the cloud and keeps the rest on the edge [13]; early-exit networks (e.g. BranchyNet [22]) aim to stop computation early to save energy. Most prior work starts computation on the client and finishes in the cloud, which makes downstream training task-dependent. We invert this order: the cloud performs heavy feature extraction and the edge executes only the final layers. This late-start design preserves the universality of a single cloud backbone while enabling cheap personalisation on low-resource devices.

Table 8.1 Frequently-used symbols

Symbol	Meaning
$\mathbf{x}, \mathbf{y}, \mathbf{z}$	Input sample, label, embedding (feature)
F	Frozen feature extractor (cloud model)
C	Lightweight classifier (edge model)
D	Original labelled *dataset* $\{(\mathbf{x}, \mathbf{y})\}$
Z	*Featureset* $\{(\mathbf{z}, \mathbf{y})\}$ derived from D
D_m, Z_m	Subset for class m ($m \in \{1, \ldots, M\}$)
$\mathcal{F}$	Embedding space $\subset \mathbb{R}^f$

8.4.5 Summary

Existing PEFT, FL, and split-inference techniques each address part of the cloud/edge efficiency challenge. Our contribution is to combine a *frozen, cloud-scale* feature extractor with *centroid-based* lightweight training, delivering privacy, adaptability, and minimal on-device computation in a single, cohesive framework.

8.5 Notation

See Table 8.1 for the definitions of frequently-used symbols.

8.6 Datasets and Featuresets

We distinguish *datasets*, consisting of raw data–label pairs $(\mathbf{x}, \mathbf{y})$, from *featuresets*, consisting of feature–label pairs $(\mathbf{z}, \mathbf{y})$. Featuresets are obtained by a single forward pass of the frozen encoder F and may include *virtual* samples that never corresponded to a real input image.

8.6.1 Combining Datasets

For disjoint class sets, multiple datasets can be merged by offsetting their label indices. For example, CIFAR-100 (M_A classes) and Flowers-102 (M_B classes) form a combined dataset D_{AB} with class indices $1 \ldots M_A$ and $M_A + 1 \ldots M_A + M_B$ respectively. The same rule applies recursively to additional datasets.

8.6.2 Full Featureset

The *full* featureset Z is simply the embedding of every sample in D; training on Z is equivalent to training on D when F is frozen (Sect. 8.3.3).

8.6.3 Class Centroid Featureset

For each class, we replace its N_m embeddings by a single virtual sample—their mean:

$$Z_{\text{avg}} = \bigcup_{m=1}^{M} \Big\{ \big(\tfrac{1}{N_m} \sum_{j=1}^{N_m} \mathbf{z}_j^m,\ m \big) \Big\}.$$

Thus $|Z_{\text{avg}}| = M$ (Fig. 8.2c).

8.6.4 Class Subcluster Centroid Featureset

To capture intra-class variation we run K-means inside each class cluster and keep the K subcluster centroids:

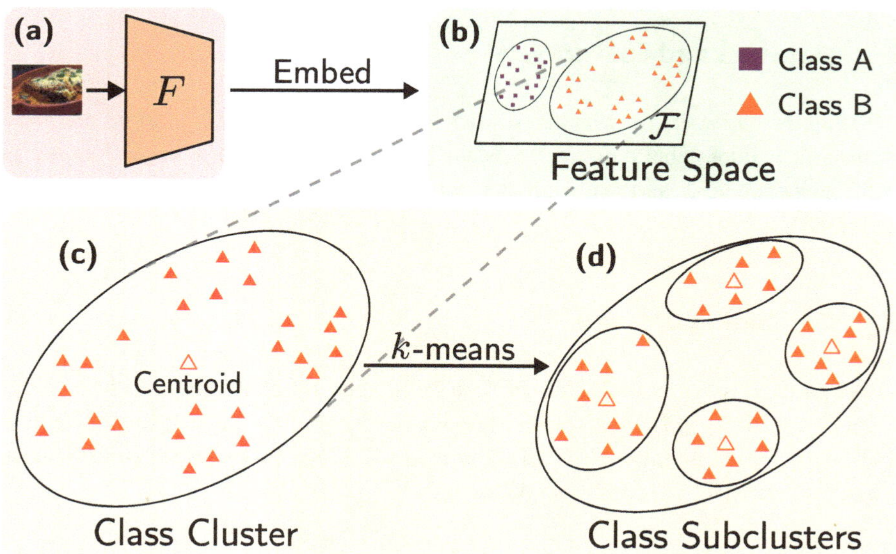

Fig. 8.2 From raw images (**a**) to embeddings (**b**) and class clusters (**c**). Running K-means within each class yields subclusters (**d**); class or subcluster centroids (empty triangles) serve as virtual samples

$$Z_{\text{sub}} = \bigcup_{m=1}^{M}\bigcup_{k=1}^{K}\Big\{\big(\tfrac{1}{N_{m,k}}\sum_{j=1}^{N_{m,k}} \mathbf{z}_j^{m,k},\ m\big)\Big\}.$$

When $K = 1$, Z_{sub} collapses to the class-centroid featureset. In our experiments we set $K = 10$.

8.7 Results

We quantify the accuracy, efficiency, and flexibility of edge-side classifiers trained on cloud-extracted embeddings. We follow the same experimental protocol for all studies unless otherwise noted.

8.8 Experimental Setup

8.8.1 Feature Extractors

We evaluate two frozen backbones:

- **CLIP** [19]: OpenCLIP ViT-B/16 with `laion2b_s34b_b88k` weights; embedding dimension $f = 512$.
- **DINOv2** [16]: ViT-B/14-reg (`dinov2_vitb14_reg`); $f = 768$.

Although both are self-supervised ViTs, CLIP is aligned to language while DINOv2 is purely visual. We discuss how these differing objectives influence centroid performance in Sect. 8.10.

8.8.2 Datasets and Featuresets

Four public datasets are used:

Dataset	#Images	#Classes
CIFAR-100 [9]	60 k	100
Food-101 [2]	101 k	101
Flowers-102 [15]	82 k	102
Tiny-ImageNet-200 [11]	110 k	200

For every dataset we create three *featuresets* (Sect. 8.6):

1. **Full:** all embeddings,
2. **Centroid:** one class centroid ($K = 1$),
3. **Subcluster:** $K = 10$ centroids per class.

8.8.3 Combined-Featureset Recipes

We denote CIFAR-100, Food-101, Flowers-102, and Tiny-ImageNet-200 by A, B, C, and D, respectively. Thus "AB" means the union of featuresets from datasets A and B.

8.8.4 Edge-Side Classifier and Training

The on-device model is a *single* linear layer (in $= f$, out $= M$). We sweep learning rates in $\{0.01, 0.05, 0.1, 0.2, 0.5\}$ using SGD with batch size 32 and train for 50 epochs (or early stop). The best learning rate is reported for each experiment. Although deeper heads could improve accuracy, they would obscure the benefit attributable to featureset design.

8.9 Evaluation

We evaluate training on the featuresets discussed in the previous section. Overall, we demonstrates the ability for local models to leverage the representations learned by ViT models for different image classification domains. In Sect. 8.9.1 we present the results of training on individual CLIP and DINOv2 featuresets. Section 8.9.2 shows we are able to achieve similar accuracy to the individual featuresets when we train on multiple featuresets using both models.

8.9.1 Individual Featuresets

Training on class centroids and subcluster centroids requires a similar number of epochs to converge compared to training on the full featuresets, as seen in Figs. 8.3 and 8.5. However, the number of samples is much lower when using class and subcluster centroids. Thus, the number of samples seen during training is a better measure for determining computation cost, and the efficiency advantage becomes much more apparent. Considering the number of samples seen, centroids and subclusters converge very rapidly, as seen in Figs. 8.4 and 8.6. As a result of this efficiency gain, hyperparameter searching is very cheap, and lends itself to the flexibility of this method for varied datasets. We also found training on

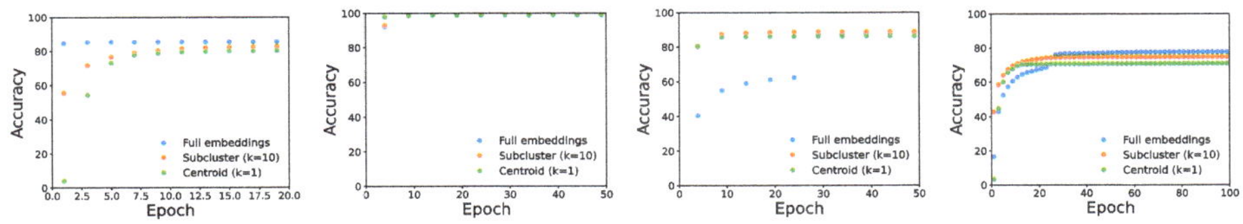

(a) CIFAR-100 (b) Flowers-102 (c) Food-101 (d) Tiny ImageNet-200

Fig. 8.3 Accuracy vs epochs when using the OpenCLIP ViT-B/16 visual encoder as the feature extractor on (**a**) CIFAR-100, (**b**) Flowers-102, (**c**) Food-101, and (**d**) Tiny-ImageNet. Performance when trained on all embeddings (blue dots) serves as the baseline. Subcluster $K = 10$ (orange) use featuresets with 10 centroids per class (i.e., subcluster centroids). Centroid $K = 1$ (green) use featuresets with only 1 centroid per class (i.e., class centroids)

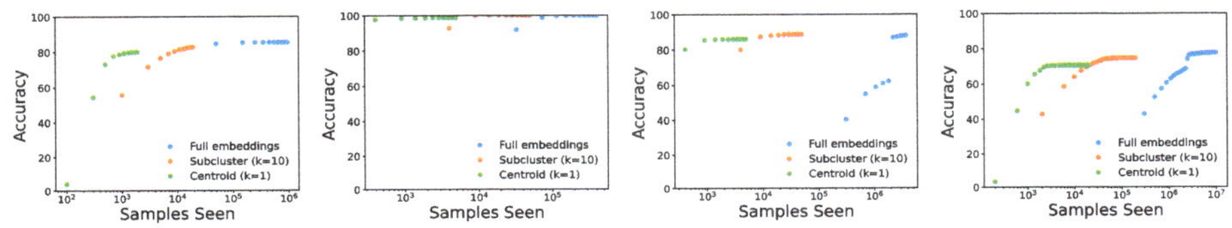

(a) CIFAR-100 (b) Flowers-102 (c) Food-101 (d) Tiny ImageNet-200

Fig. 8.4 Accuracy vs number of samples seen when using the OpenCLIP ViT-B/16 visual encoder as the feature extractor on (**a**) CIFAR-100, (**b**) Flowers-102, (**c**) Food-101, and (**d**) Tiny-ImageNet. Performance when trained on all embeddings (blue dots) serves as the baseline. Subcluster $K = 10$ (orange) use featuresets with 10 centroids per class (i.e., subcluster centroids). Centroid $K = 1$ (green) use featuresets with only 1 centroid per class (i.e., class centroids). Using class centroids or subcluster centroids significantly reduces the number of samples required to achieve full accuracy, and centroids offer respectable performance

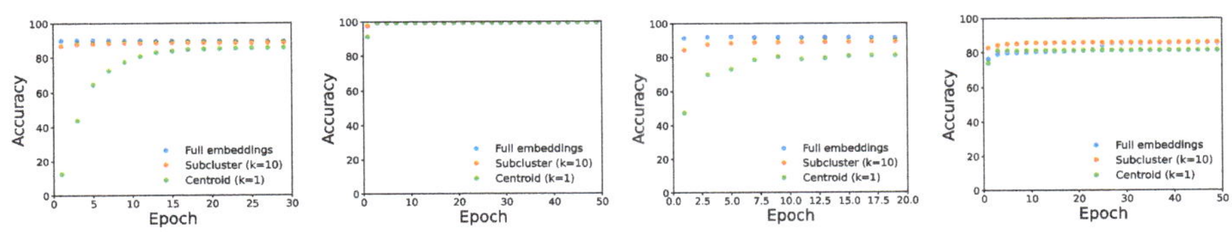

(a) CIFAR-100 (b) Flowers-102 (c) Food-101 (d) Tiny ImageNet-200

Fig. 8.5 Accuracy vs epochs when using the DINOv2 ViT-B/14 model as the feature extractor on (**a**) CIFAR-100, (**b**) Flowers-102, (**c**) Food-101, and (**d**) Tiny-ImageNet. Performance when trained on all embeddings (blue dots) serves as the baseline. Subcluster $K = 10$ (orange) use featuresets with 10 centroids per class (i.e., subcluster centroids). Centroid $K = 1$ (green) use featuresets with only 1 centroid per class (i.e., class centroids)

featuresets with fewer samples required higher learning rates (i.e., generally, learning rate for centroids > subclusters > full featuresets).

Our results show that training on virtual samples achieves competitive accuracy with only a fraction of the samples seen. This makes retraining, e.g., for a new classifier head, significantly more feasible on smaller devices.

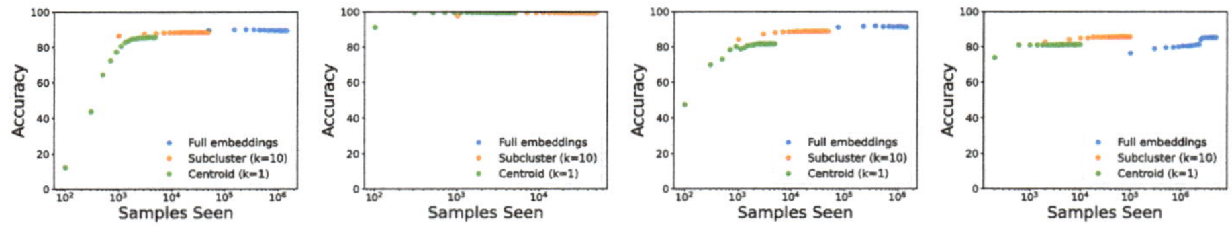

(a) CIFAR-100 (b) Flowers-102 (c) Food-101 (d) Tiny ImageNet-200

Fig. 8.6 Accuracy vs number of samples seen when using the DINOv2 ViT-B/14 model as the feature extractor on (**a**) CIFAR-100, (**b**) Flowers-102, (**c**) Food-101, and (**d**) Tiny-ImageNet. Performance when trained on all embeddings (blue dots) serves as the baseline. Subcluster $K = 10$ (orange) use featuresets with 10 centroids per class (i.e., subcluster centroids). Centroid $K = 1$ (green) use featuresets with only 1 centroid per class (i.e., class centroids). Using class centroids or subcluster centroids significantly reduces the number of samples required to achieve full accuracy, and centroids offer respectable performance

Figure 8.3 shows the accuracy for training with CLIP-derived features. We see that all three featuresets converge to roughly the same accuracy. As expected, the full embeddings featureset performs slightly better (1–5%) than the centroid-based featuresets. Figure 8.4 demonstrates how the centroid-based featuresets require significantly less compute when compared to the full dataset.

Figures 8.5 and 8.6 show the same results but with the DINOv2 model as the feature extractor. Overall accuracy for DINOv2 is slightly higher than that of CLIP. We suspect the larger embedding dimension $f = 768$ may help with the classifier's discriminatory capabilities.

8.9.2 Combined Featuresets

Creating combined featuresets with centroids from multiple datasets does not significantly affect accuracy on individual datasets, as seen in Figs. 8.7 and 8.8. However, using a new featureset on an already trained classifier leads to rapid forgetting, getting close to zero percent accuracy after a few epochs. We suspect that since the classifer sees no examples of a class it sets the probability of it to zero.

For our experiments, we make the simplifying assumption the classes are disjoint to make indexing easier for the combining method described in Sect. 8.6.1. However, we acknowledge that there are overlaps: e.g., "rose" is present in both CIFAR-100 and Flowers-102; "pizza" is present in both CIFAR-100 and Food-101. No additional processing was done to merge these labels or remove data with overlapping labels. We suspect this may be the reason for minor accuracy degradation when using DINOv2 features (Fig. 8.8).

We also trained the using the subcluster featuresets (Figs. 8.9 and 8.10). We observed results similar to the individual featureset training in Sect. 8.9.1: subclusters attained slightly better accuracy than centroids.

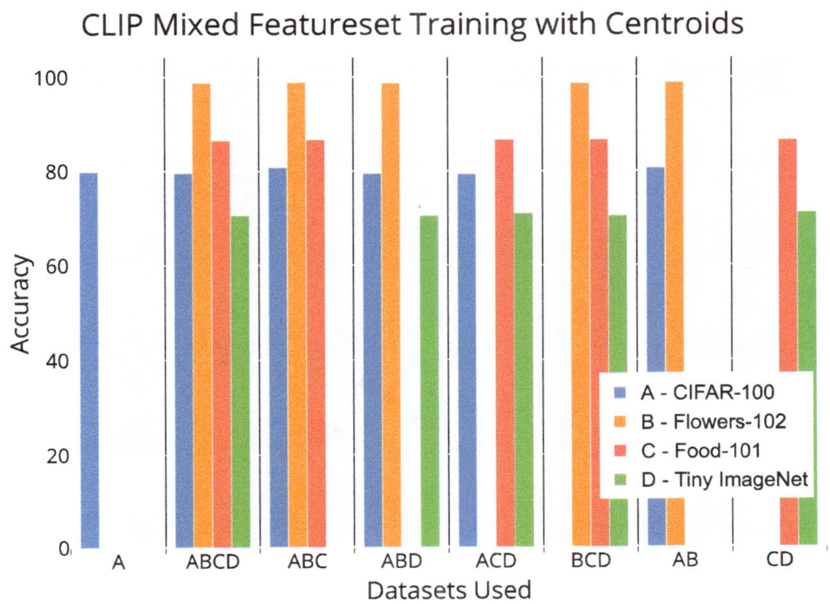

Fig. 8.7 Test accuracies using featuresets comprising CLIP-derived centroids from different datasets. The x-axis represents the recipe (e.g., AB is a combined featureset with CIFAR-100 and Flowers-102 centroids), and the y-axis represents accuracy. Individual bars are evaluated test accuracies for different datasets. We observe that combining featuresets does not impact classifier accuracy when evaluated on individual datasets

8.10 Discussion

Why Do Centroids Work? For CLIP, mutual visual–text alignment places semantically similar images close in cosine space [19]; a centroid is therefore a meaningful prototype. DINOv2 lacks text supervision, yet its self-supervised objective still organizes images into visually coherent clusters, explaining the strong centroid performance—consistent with findings on grounded pre-training [12].

Privacy and Deployment Labels stay on-device; raw images can optionally be discarded once embeddings are cached. Where stronger privacy is required, an on-device emulator of the early backbone layers [23] could avoid sending any raw data to the cloud. Model-splitting does introduce a potential attack surface for feature inversion [3], which merits future study.

Feature Augmentation Simple Gaussian noise on embeddings did not improve accuracy; augmenting in feature space remains an open problem.

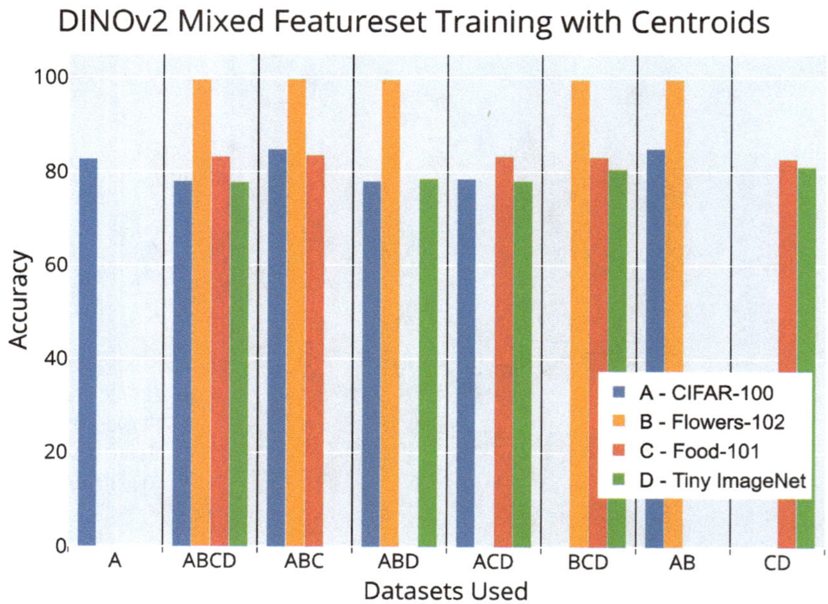

Fig. 8.8 Test accuracies using featuresets comprising DINOv2-derived centroids from different datasets. The x-axis represents the recipe (e.g., AB is a combined featureset with CIFAR-100 and Flowers-102 centroids), and the y-axis represents accuracy. Individual bars are evaluated test accuracies for different datasets. Adding additional datasets (e.g., A → AB) DINOv2 featuresets slightly impacts performance on individual datasets (e.g., compare A, ABD, ABCD; and ABCD, CD)

8.11 Summary

This chapter has presented a complete pipeline for *leveraging cloud knowledge locally*: a large, frozen feature extractor in the cloud and an ultra-lightweight classifier that is trained and executed entirely on the edge device. Our main findings are:

- **Embeddings as a universal interface.** Self-supervised backbones such as CLIP and DINOv2 generate rich visual embeddings that generalize across tasks and datasets. These embeddings can be transmitted once and reused indefinitely.
- **Virtual samples enable extreme data reduction.** Replacing thousands of embeddings per class with a single *centroid* preserves accuracy within 1–5 pp of the full featureset, while cutting storage and training time by two orders of magnitude.
- **Rapid, private personalisation.** Because only the classifier head is updated, a new class can be added—or an entire head re-trained—in seconds on commodity hardware, without sharing labels or raw images with the cloud.

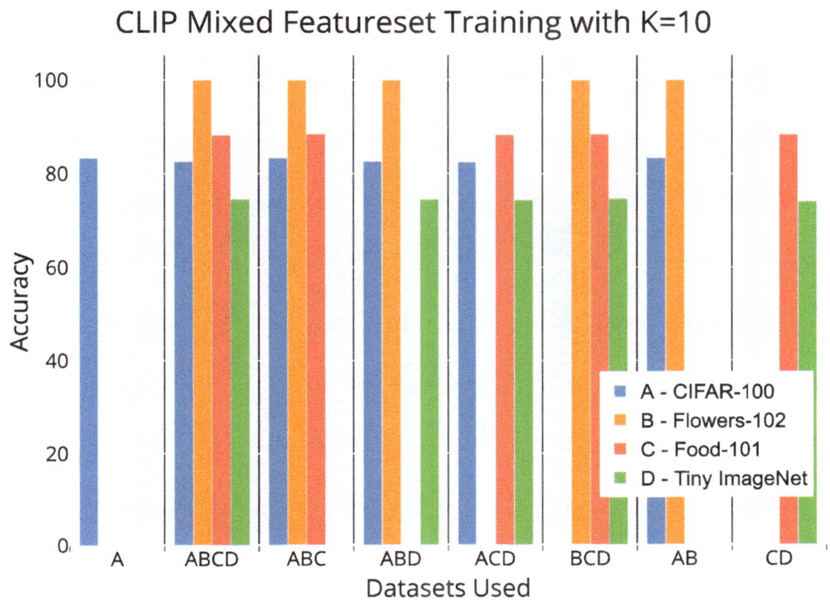

Fig. 8.9 Test accuracies using featuresets comprising CLIP-derived subclusters from different datasets. The x-axis represents the recipe (e.g., AB is a combined featureset with CIFAR-100 and Flowers-102 subclusters), and the y-axis represents accuracy. Individual bars are evaluated test accuracies for different datasets. We observe that adding additional featuresets (e.g., A → AB) does not impact classifier accuracy when evaluated on individual datasets (bars for A and B)

- **Composability across domains.** Centroids drawn from multiple, disjoint datasets can be merged into a single featureset with negligible loss in per-dataset accuracy, enabling multi-domain edge models with no additional cloud retraining.

Implications The split architecture combines the expressive power of foundation-scale backbones with the privacy, latency, and energy benefits of on-device inference. It removes the requirement to store or execute the full model locally.

Future Directions Key open questions include (i) adaptive centroid selection for evolving data distributions; (ii) privacy guarantees against feature inversion; (iii) extension to multi-modal backbones such as ImageBind; and (iv) integration with federated learning, where only the tiny heads are aggregated.

Overall, the proposed framework introduces a practical approach for the deployment of sophisticated visual intelligence on resource-constrained devices.

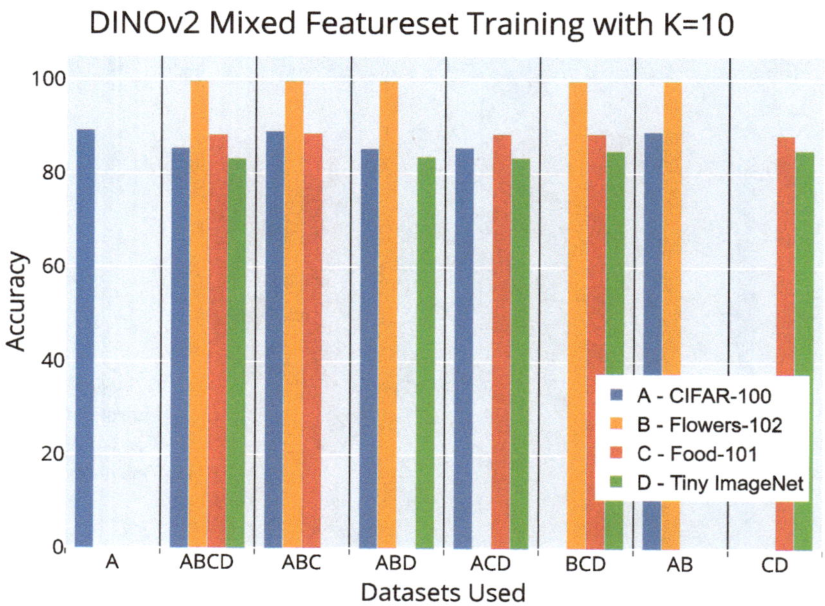

Fig. 8.10 Test accuracies using featuresets comprising DINOv2-derived subclusters from different datasets. The x-axis represents the recipe (e.g., AB is a combined featureset with CIFAR-100 and Flowers-102 subclusters), and the y-axis represents accuracy. Individual bars are evaluated test accuracies for different datasets. Adding additional datasets (e.g., A → AB) DINOv2 featuresets slightly impacts performance on individual datasets (e.g., compare A, ABCD, AB; and BCD, CD)

References

1. Guillaume Alain and Yoshua Bengio. Understanding intermediate layers using linear classifier probes, November 2018.
2. Lukas Bossard, Matthieu Guillaumin, and Luc Van Gool. Food-101 – Mining Discriminative Components with Random Forests. In David Fleet, Tomas Pajdla, Bernt Schiele, and Tinne Tuytelaars, editors, *Computer Vision – ECCV 2014*, volume 8694, pages 446–461. Springer International Publishing, Cham, 2014.
3. Xin Dong, Hongxu Yin, Jose M. Alvarez, Jan Kautz, Pavlo Molchanov, and H. T. Kung. Privacy Vulnerability of Split Computing to Data-Free Model Inversion Attacks, October 2022.
4. Xin Dong, Sai Qian Zhang, Ang Li, and H.T. Kung. SphereFed: Hyperspherical Federated Learning. In Shai Avidan, Gabriel Brostow, Moustapha Cissé, Giovanni Maria Farinella, and Tal Hassner, editors, *Computer Vision – ECCV 2022*, volume 13686, pages 165–184. Springer Nature Switzerland, Cham, 2022.
5. Alexey Dosovitskiy, Lucas Beyer, Alexander Kolesnikov, Dirk Weissenborn, Xiaohua Zhai, Thomas Unterthiner, Mostafa Dehghani, Matthias Minderer, Georg Heigold, Sylvain Gelly, Jakob Uszkoreit, and Neil Houlsby. An image is worth 16x16 words: Transformers for image recognition at scale, 2021.

6. Edward J. Hu, Yelong Shen, Phillip Wallis, Zeyuan Allen-Zhu, Yuanzhi Li, Shean Wang, Lu Wang, and Weizhu Chen. LoRA: Low-Rank Adaptation of Large Language Models, October 2021.
7. Simon Kornblith, Mohammad Norouzi, Honglak Lee, and Geoffrey Hinton. Similarity of neural network representations revisited, 2019.
8. Simon Kornblith, Mohammad Norouzi, Honglak Lee, and Geoffrey Hinton. Similarity of Neural Network Representations Revisited. In Kamalika Chaudhuri and Ruslan Salakhutdinov, editors, *Proceedings of the 36th International Conference on Machine Learning*, volume 97 of *Proceedings of Machine Learning Research*, pages 3519–3529. PMLR, June 2019.
9. Alex Krizhevsky. Learning multiple layers of features from tiny images. Technical report, University of Toronto, 2009.
10. Ananya Kumar, Aditi Raghunathan, Robbie Matthew Jones, Tengyu Ma, and Percy Liang. Fine-Tuning can Distort Pretrained Features and Underperform Out-of-Distribution. In *International Conference on Learning Representations*, October 2021.
11. Ya Le and Xuan Yang. Tiny imagenet visual recognition challenge. Stanford CS231n Course Project, 2015.
12. Shilong Liu, Zhaoyang Zeng, Tianhe Ren, Feng Li, Hao Zhang, Jie Yang, Chunyuan Li, Jianwei Yang, Hang Su, Jun Zhu, and Lei Zhang. Grounding DINO: Marrying DINO with Grounded Pre-Training for Open-Set Object Detection, March 2023.
13. Yoshitomo Matsubara, Davide Callegaro, Sabur Baidya, Marco Levorato, and Sameer Singh. Head Network Distillation: Splitting Distilled Deep Neural Networks for Resource-Constrained Edge Computing Systems. *IEEE Access*, 8:212177–212193, 2020.
14. Brendan McMahan, Eider Moore, Daniel Ramage, Seth Hampson, and Blaise Aguera y Arcas. Communication-Efficient Learning of Deep Networks from Decentralized Data. In *Proceedings of the 20th International Conference on Artificial Intelligence and Statistics*, pages 1273–1282. PMLR, April 2017.
15. Maria-Elena Nilsback and Andrew Zisserman. Automated Flower Classification over a Large Number of Classes. In *2008 Sixth Indian Conference on Computer Vision, Graphics & Image Processing*, pages 722–729, December 2008.
16. Maxime Oquab, Timothée Darcet, Théo Moutakanni, Huy Vo, Marc Szafraniec, Vasil Khalidov, Pierre Fernandez, Daniel Haziza, Francisco Massa, Alaaeldin El-Nouby, Mahmoud Assran, Nicolas Ballas, Wojciech Galuba, Russell Howes, Po-Yao Huang, Shang-Wen Li, Ishan Misra, Michael Rabbat, Vasu Sharma, Gabriel Synnaeve, Hu Xu, Hervé Jegou, Julien Mairal, Patrick Labatut, Armand Joulin, and Piotr Bojanowski. DINOv2: Learning Robust Visual Features without Supervision, April 2023.
17. Sinno Jialin Pan and Qiang Yang. A Survey on Transfer Learning. *IEEE Transactions on Knowledge and Data Engineering*, 22(10):1345–1359, October 2010.
18. Jonas Pfeiffer, Aishwarya Kamath, Andreas Rücklé, Kyunghyun Cho, and Iryna Gurevych. AdapterFusion: Non-Destructive Task Composition for Transfer Learning. In *Proceedings of the 16th Conference of the European Chapter of the Association for Computational Linguistics: Main Volume*, pages 487–503, Online, 2021. Association for Computational Linguistics.
19. Alec Radford, Jong Wook Kim, Chris Hallacy, Aditya Ramesh, Gabriel Goh, Sandhini Agarwal, Girish Sastry, Amanda Askell, Pamela Mishkin, Jack Clark, Gretchen Krueger, and Ilya Sutskever. Learning Transferable Visual Models From Natural Language Supervision, February 2021.
20. Virginia Smith, Chao-Kai Chiang, Maziar Sanjabi, and Ameet Talwalkar. Federated multi-task learning. In *Proceedings of the 31st International Conference on Neural Information Processing Systems*, NIPS'17, pages 4427–4437, Red Hook, NY, USA, December 2017. Curran Associates Inc.

21. Chuanqi Tan, Fuchun Sun, Tao Kong, Wenchang Zhang, Chao Yang, and Chunfang Liu. A Survey on Deep Transfer Learning. In Věra Kůrková, Yannis Manolopoulos, Barbara Hammer, Lazaros Iliadis, and Ilias Maglogiannis, editors, *Artificial Neural Networks and Machine Learning – ICANN 2018*, Lecture Notes in Computer Science, pages 270–279, Cham, 2018. Springer International Publishing.
22. Surat Teerapittayanon, Bradley McDanel, and H. T. Kung. BranchyNet: Fast Inference via Early Exiting from Deep Neural Networks, September 2017.
23. Guangxuan Xiao, Ji Lin, and Song Han. Offsite-Tuning: Transfer Learning without Full Model, February 2023.
24. Liangjun Xie, Rui Huang, Nong Gu, and Zhiqiang Cao. A novel defect detection and identification method in optical inspection. *Neural Computing and Applications*, 24(7):1953–1962, June 2014.
25. Chongzhi Zhang, Mingyuan Zhang, Shanghang Zhang, Daisheng Jin, Qiang Zhou, Zhongang Cai, Haiyu Zhao, Xianglong Liu, and Ziwei Liu. Delving Deep Into the Generalization of Vision Transformers Under Distribution Shifts. In *Proceedings of the IEEE/CVF Conference on Computer Vision and Pattern Recognition*, pages 7277–7286, 2022.
26. Ce Zhou, Qian Li, Chen Li, Jun Yu, Yixin Liu, Guangjing Wang, Kai Zhang, Cheng Ji, Qiben Yan, Lifang He, Hao Peng, Jianxin Li, Jia Wu, Ziwei Liu, Pengtao Xie, Caiming Xiong, Jian Pei, Philip S. Yu, and Lichao Sun. A Comprehensive Survey on Pretrained Foundation Models: A History from BERT to ChatGPT, May 2023.

Conclusion

9

9.1 A Unified Matrix-Centric Framework

This book has argued that many of the core computational challenges in modern machine learning can be addressed by optimizing the use of matrices. By developing a set of matrix-centric techniques, we have shown how to improve the efficiency, scalability, and sustainability of artificial intelligence workloads by changing how matrices are stored, accessed, and compressed.

Throughout this work, hardware resource constraints served as the central motivator. Whether in edge devices, datacenters, or distributed systems, the continual push toward larger and more capable models repeatedly runs into limits on computation, memory bandwidth, local storage, or communication cost. Alleviating one bottleneck often uncovers another; however, the frameworks introduced here provide the analytical tools to navigate these trade-offs.

9.1.1 Summary of Contributions

Optimal Tiling and Scheduling The *CAKE* framework (Chap. 2) demonstrated how strategic tiling and block shaping can mitigate external memory bandwidth constraints. By using an analytical model to reason about the trade-offs between local memory usage and bandwidth, we circumvented the need for extensive autotuning. This approach was further generalized in *mCAKE* (Chap. 3) to address higher-dimensional tensor contractions, demonstrating that arithmetic intensity analysis can yield high-performing, hardware-agnostic solutions.

A. Sabot, *Minimizing Data Movement and Parameter Count Across the Machine Learning Stack*, Synthesis Lectures on Computer Science,
https://doi.org/10.1007/978-3-032-23100-0_9

Hardware-Aware Sparsity As AI models grow and their matrices become sparse, skipping zero elements often disrupts the balance between computation and data transfer. The *Rosko* framework (Chap. 4) addressed this through a hybrid training pipeline, allowing neural networks to be pruned and fine-tuned into sparsity patterns directly compatible with hardware memory access sizes.

Structural Compression and Distillation When sparsity alone is insufficient, low-rank matrix approximation via Singular Value Decomposition (SVD) offers a viable alternative. We introduced new algorithms for efficient inference with compressed weights (Chap. 5) and established a *two-stage adaptation* strategy to maintain accuracy during fine-tuning (Chap. 6). To bridge the performance gap in highly compressed models, the *block-by-block knowledge distillation* technique (Chap. 7) was introduced, allowing student networks to learn from teacher counterparts at a modular level, significantly reducing the hardware resources required for training.

Collaborative and Split Computing Finally, we extended these optimizations to the realm of split computing (Chap. 8). By partitioning models between local edge devices and the cloud, we demonstrated how to leverage massive "cloud knowledge" while preserving local privacy and enabling rapid personalization.

9.2 The Path Forward: Future Directions

The methodologies explored here are well-positioned to guide optimization in distributed and hierarchical systems, where communication bottlenecks manifest at every level between large, slow memories and fast, local buffers.

As deep learning models continue to expand, the techniques for training, compressing, and efficiently storing weights will become even more critical. One emerging frontier is the management of intermediate states, such as **KV caches** in large language models. These structures represent a massive and growing portion of the memory footprint in modern inference, and applying the matrix-centric principles of bandwidth-aware scheduling and low-rank compression to these dynamic structures is a promising avenue for research.

In summary, this book demonstrates that an analytical understanding of matrix operations enables the scalable, hardware-aware, and sustainable deployment of AI. While new bottlenecks will inevitably arise, these results offer durable foundations on which the next generation of efficient machine learning can be built.

Index

A. Sabot, *Minimizing Data Movement and Parameter Count Across the Machine Learning Stack*, Synthesis Lectures on Computer Science,
https://doi.org/10.1007/978-3-032-23100-0

MIX
Papier | Fördert gute Waldnutzung
FSC® C083411

Zeitfracht Medien GmbH
Ferdinand-Jühlke-Straße 7
99095 Erfurt, Deutschland
produktsicherheit@kolibri360.de